HYDROPONIC

ORGANIC

GARDENING

NICK BARAS

INTRODUCTION

Many amateur gardeners and householders would like to take up hydroponics if only they knew how to set about it. Others may be equally anxious to begin growing plants without soil but perhaps feel some diffidence in starting on a new system in the absence of clear and well-defined guidance. Possibly they have heard, or been told—quite inaccurately as it happens—that soilless gardening is rather complicated and requires specialized knowledge.

This idea that hydroponics always needs a very high standard of technical ability to make it succeed has discouraged large numbers of people from even trying it out. But today it exists, at least one straightforward method of gardening without soil that can be used with complete confidence by ordinary gardeners, homemakers, and indeed anyone prepared to follow a few basic and easily understood rules.

Those people who have started simplified hydroponics have been delighted with the results that they have obtained and the excellent quality of the flowers and vegetables that they have grown so successfully in such short periods, without any particular effort or expense.

So why not you? This book details an effortless way of growing your household supplies of garden flowers and green food. No soil at all is required.

The beginner will be introduced step by step, to hydroponics and shown how to set up and operate easy-to-use soilless gardens in the home or backyard. The method recommended has been tested by long experience and found to be well suited to the needs of amateurs and homemakers. Clear and concise instructions are given about daily work and the care of hydroponic units.

Hydroponics can be an enjoyable hobby, a profitable pastime, or a way of adding to your income by growing your fresh produce at home. It will also guide you towards a better understanding of Nature and a more profound knowledge of biological science. Gardening is dear to the hearts of most of us, and in the soilless culture, you will surely find the means to keep in close and regular contact with plant life. Added to all, this is the fact that hydroponics demands no hard work or long hours of labor.

Begin with the simple method recommended in this book. Once you have become familiar with it, branch out, if you wish, into some of the more advanced techniques discussed in the final chapter. But no matter how small your soilless garden may be—or remain— you will always find something to delight you in the ease and simplicity of amateur hydroponics

TABLE OF CONTENTS

A DEFINITION OF HYDROPONICS

Hydroponics is generally the science of growing plants without using soil by feeding them on solutions of water and mineral salts instead of relying upon the traditional methods of cultivating the earth, which most gardeners and farmers still continue to do in order to raise crops.

However, because such a description could sound a little complicated, let us begin by thinking in simpler terms.

It is not always realized that hydroponics might be employed in numerous ways, from the large-scale production of commercial foodstuffs, flowers, and fruits, through medium-sized office and community growing units, right down to smaller lots of home blooms and indoor plants, or tasty vegetables and succulent salads for the household, or as a family hobby.

This striking versatility of soilless culture, combined with the excellent results that are obtained in all kinds of places, has made the system ideal for widely differing conditions

Hydroponic gardening can be more comfortable and more pleasant to do than conventional soil gardening.

During the past few years, an immense amount of scientific research has been undertaken to develop effortless and practical ways of growing plants without soil, methods that can be used confidently by amateurs and homemakers.

So do not be put off by something you have perhaps read about hydroponics being a complicated procedure requiring specialized skills or knowledge for success. On the contrary, simplified gardening without soil is straightforward if you follow the few easy rules. Naturally, in big commercial units, there has to be technical control, but for growing plants in the home or ordinary garden, backyard or window ledge, what matters most are love, care, and ability to stick to the guidelines laid down.

The term hydroponics derives from two Greek words, *hudor,* water, and *ponos* work.

When combined, these mean *'water-working,'* and a reference to the use of solutions of water and fertilizer chemicals for soilless plant cultivation, as opposed to usual growth in soil or *geoponics* (the care of the earth).

Quite many different methods of hydroponics are today in general usage throughout the world, the exact choice depending upon local needs and conditions. However, they all conform to the same basic principles, and all have one common objective: the growing of plants quite independently of earth or organic matter.

This capacity for divorcing the cultivation of flowers, vegetables, fruits, and grains from the soil has profound implications for the development of horticulture and farming.

In short, it means that humanity is no longer solely dependent on the land for sustenance and the satisfaction of natural desires for a beautiful environment because gardeners and farmers can now produce lovely flowers and raise beautiful crops without any soil in almost any place they may wish.

Given a minimum supply of water, hydroponic units, both small and large, may be set up in towns and cities, for the enjoyment and use of urban populations.

At the same time, people living in deserts or barren regions can raise quantities of attractive and healthy plants even under the very sterile conditions common in such areas.

Apart from these general considerations, hydroponics also offers certain essential advantages to the individual gardener or householder:

better quality plants, quicker growth, much saving of time and labor, the elimination of troublesome tasks like digging, weeding and manuring, lower costs, a complete absence of dirt and smells remarkably consistent results, and attractive year-round displays of blooms or quantities of fresh greenstuff in the home.

HOW PLANTS GROW

Although the cycle of plant life is complicated stuff, it can be explained in simple terms. A plant is a sort of natural workshop, each section of which is engaged in the task of changing raw food material into living tissue. To ensure proper growth, all higher or green plants, and the vast majority of flowers and vegetables belong to this category, require certain essential things, including water, light, air, mineral salts, and support for the roots.

Unlike animals and human beings, most plants cannot ingest solid or organic food material, so they are obliged to absorb part of their nourishment from gases in the air and part from solutions of inorganic salts or chemicals and water.

These pure substances are transformed by the various departments of the plants into living tissue through the expenditure of energy obtained from light. Some oxygen is needed for these processes, but a significant contribution to growth and development is made by the gas called carbon dioxide present in the atmosphere.

To do the job, mineral salts, in combination with water, are absorbed through tiny hairs located on the plants' roots, by the use of a force known as osmosis. Plants always grow upwards towards the light.

To permit them to stand, adequate support is vital. In soil, the earth provides this, but in hydroponics, it is necessary to supply alternative devices or growing media.

It is essential to choose the right kinds of supports because roots need to breathe just as leaves and stems do.

At the same time, there has to be enough moisture around them to prevent them from drying off or dying from lack of water and food-air (carbon dioxide and oxygen)

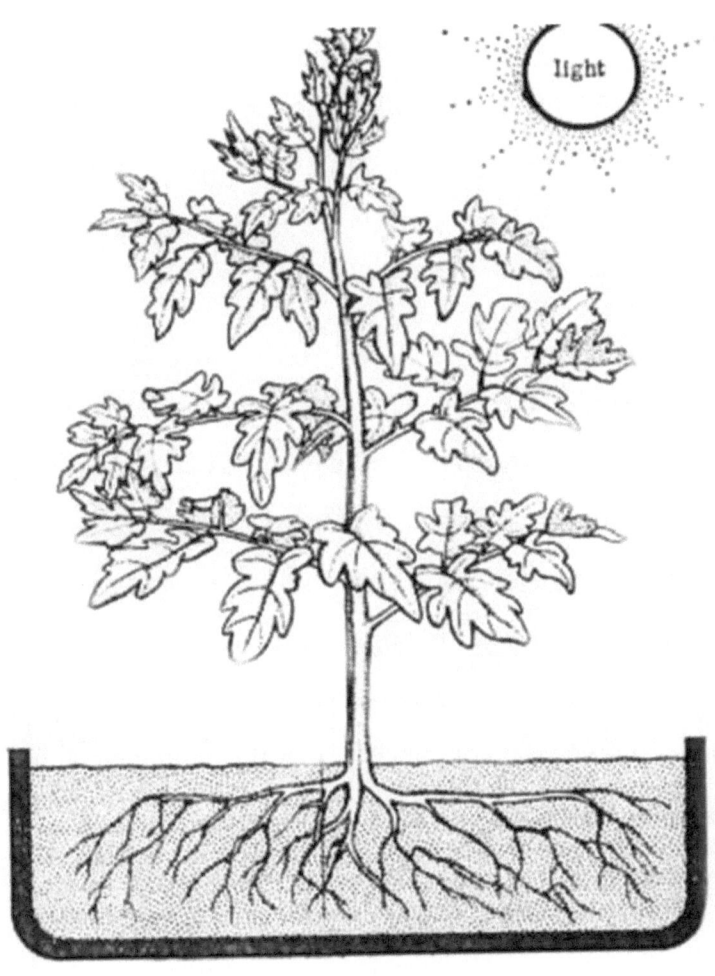

Here we can see how the plant's essential needs are provided without the use of soil.

Nature gives us light, air, and water, or we can provide them ourselves. At the same time, instead of earth and manures, we substitute an aggregate to support the roots and a solution of water and fertilizer salts to furnish adequate foodstuff. Warmth can come from sunshine or different methods of heating.

Now, suppose we take a dry seed and watch its behavior and the different functions of the various parts during development until it becomes a flowering plant. The easiest way to do this is to bury the seed in a little moist sawdust. Soon, germination will occur, and the seed will have absorbed enough water so that the outer skin bursts and the tiny embryo plant, which is contained within the seed coat, starts to develop.

The young root, called the radicle, forces its way downwards to grip firmly the support, in our case, the sawdust, while a small, pointed stem, termed the plumule, grows upwards towards the light.

So far, the young plant has developed by feeding on food material stored within the seed and by using the essentials that we provided for it, water, air, and support for the roots. Light also is present and available as soon the plumule pushes its way up into the open air since we have taken care to let sunshine or other illumination reach it daily.

But shortly, the growing plant will have another demand to make, this time for mineral salts. If these are not supplied, the seedling will quickly weaken and die. Consequently, we must feed our plant regularly with some suitable nutrients. If this is done, nothing, except disease or accident, will prevent our seedling from developing into a healthy and mature plant.

THE HYDROPONIC PROCESS

Of the five essential requirements for good plant growth, three are supplied in hydroponics generally by the same means as they are in soil gardening.

Water, light, and air are in Nature's gift and may be obtained in the home as part of our surroundings or through the ingenuity of man.

The last two, mineral salts and support for the root, must be procured as extras. To cultivate the earth, we dig and hoe and add manures or composts. These latter are organic materials, but plants cannot use these directly.

They have to go through a lengthy process of weathering or breaking down in the ground before they become available as inorganic nutrients. Even then, the gardener or farmer has no real idea of just how much plant food may be present in the land at any given time. In hydroponics, on the other hand, the application of nutrients is a controlled and balanced procedure.

We know that some eleven different elements are, in the main, necessary for proper growth, in addition to the oxygen, these are nitrogen, phosphorus, potassium, calcium, magnesium, sulfur, iron, manganese, boron, zinc, and copper.

The soilless gardener, therefore, dispenses with the laborious and often back-breaking task of working the land and instead employs a carefully prepared nutrient mixture or formula to feed plants. This will act quickly and immediately on application, being absorbed without delay in solution with water through the root hairs.

For practical purposes, in-home or commercial hydroponics, we always use fertilizer-type chemicals as the sources of nutrients.

These are cheaper and easier to buy and are readily available in a convenient form. Root supports have already been mentioned. There is quite a wide choice of materials available for soilless gardeners. Unlike earth, which contains various substances, aggregates or growing media are virtually inert. They cannot be damaged by the elements and are not subject to rapid erosion as the soil is.

The primary function of the supports is to provide an anchor for plants' roots. In addition, they also serve as a reservoir for the nutrient solution of water and fertilizer salts, which contain the vital foodstuffs for growth.t

The main differences between hydroponics and soil growing should now be clear to the reader. Not only is the soilless garden exact and controlled in its functioning, but it also cuts out many time-consuming and laborious jobs necessary in conventional plant cultivation. Messy manures, often smelly and slow-acting, are replaced by clean, swiftly available, mineral nutrients, and the frequently unreliable earth or expensive and troublesome composts are entirely eliminated in favor of an easily handled aggregate, or another supporting device.

These advantages, together with many others already discussed, are what has made so many lovers of plants turn to hydroponics as the ideal system for use in the home.

EXAMPLE OF GROWING UNITS

Soilless garden units for simple household hydroponics are easy to set up. The method recommended here for beginners is scientifically designed to give good results. At the same time, every effort has been made to avoid any complicated equipment which home gardeners or housewives might find a little challenging to use. Individual preferences and local needs often vary very much, and therefore the final choice of containers or receptacles for plants may well be left to personal taste and convenience.

A hydroponic unit is made up of the following parts:

- A container, also called a bed or trough, though frequently pots and miscellaneous receptacles are employed;

- The aggregate or growing medium, which is placed in the container and provides support for the roots;

- Water supply. In most installations, the plant food, in the form of fertilizer salts, is added to the water to constitute the nutrient solution.

This is then applied to the beds, troughs or pots in regular watering. To work the soilless garden, you will also need a few items of the standard pattern, such as a can, a small hand fork, a kitchen balance, and some string or twine to tie up taller plants. In this chapter, we will discuss the different parts of the home hydroponic unit, how to set them up, and suggest some positions for plants growing in and around the house.

CONTAINERS

Various kinds of containers may be used for home growing without soil. Shallow boxes, pots, bowls, old kitchen sinks, and other receptacles are entirely satisfactory.

It is not difficult to construct troughs of any desired size from bricks and mortar or concrete.

Usually, a depth of six inches is best. This allows adequate room for root development and is not too tall.

Hydroponic plants, in fact, require less space or depth for rooting than do soil-grown ones, because their roots are more compact and have ample supplies of essential nutrients immediately available.

Containers should not be made from any material likely to prove toxic to plants. If of galvanized iron sheeting, then they should be painted with a good quality varnish or paint before use. Wooden containers may be lined.

Polythene, which is waterproof and can be fixed in position with drawing-pins. Ready-made plastic troughs of various types are available in shops.

Other suggestions include cut-down barrels and cans, OR oil drums sawn in sections.

The length and width of soilless gardens will depend on the needs of the householder, but generally speaking, it is desirable to keep the width of units to three feet or under, because otherwise, it becomes difficult to attend appropriately to the plants growing in them. Any convenient length may be employed to suit particular situations.

It is essential to be sure that containers for hydroponics have holes in the base. In fact, only one hole of a diameter not exceeding a quarter to a half of an inch is needed in pots; in troughs, however, several holes may be desirable.

If there are too many apertures, close them up with Plasticine, screwed-up wads of newspaper or wooden plugs, leaving only enough for drainage and aeration.

Most plant pots, as well as some troughs offered for sale, do have holes already punched in the bottom, or else a few marks in the bases showing where you can push a pencil or skewer through to form a suitable aperture. In hydroponic gardening, the holes are only opened at certain times and are fitted with small plugs or stoppers. To catch any seepage, shallow dishes, saucers, or old tin lids can be placed underneath pots and gutters fixed beside troughs, beds, or more prolonged containers.

To make an elementary room or kitchen trough, ask your greengrocer to give you a light wooden box of the kind used for packing wooden box lining of polythene sheeting drainage holes with removable plugs

Grapes, pears, or large peaches.

This will be about two feet long by one to one-and-a-half feet wide and some six inches deep.

Box sizes do vary, but these measurements are reasonably common ones. Place the box on a table or a similar stand. Then take some polythene plastic sheeting, or opened bags of this material, like those used for wrapping different bulky goods.

Line the box carefully with the waterproof plastic, turning it up at the sides and ends and fixing it in position with drawing pins to the top edges of the box. Now, using a gimlet or sharp skewer, make small holes half-an-inch from the bottom of the box, two on each side, and two at one end only. Put two small stones or a strip of thin wood under the other end, so as to raise it not more than three-quarters of an inch. Finally, put little plugs in each of the holes you made. Your hydroponic trough or container will now be ready and can be filled with aggregate. It is quite easy to make lots of such simple units, arranging them in series or as may be convenient.

AGGREGATES

Once you have chosen your containers or made them from available materials, the next job is to fill them with aggregate or growing medium. Quite a number of different aggregates are satisfactory for hydroponics. Very often, however, it may be most economical to use a growing medium which is in plentiful supply locally and will cost little or nothing. For example, sand, gravel, or small pebbles can frequently be gathered near the home, without the expense, while cinders from coal fires only require washing to make them suitable for soilless gardening. Here is a list of common aggregates employed by hydroponics.

Sand: The best sands for hydroponic gardening are relatively coarse ones, especially river or beach sands. Too fine sands are liable to become waterlogged, thus preventing proper aeration of the plants' roots. In technical terms, we classify sand size on the basis of mesh, that is to say, sieve measurement. A sieve size range of from 14 to 100 mesh is highly satisfactory, consisting of several grades. Sands with a uniform particle diameter of under 30 mesh should not be used alone but should be mixed with a proportion of coarser materials. As an example, In Aruba, coarser sands have been employed with much success, even those having a particle size of one-sixteenth to one-eighth of an inch. For home growing purposes, under most conditions, sands should not contain more than half of their volume of material below 30 mesh sieve size, with the balance made up of larger grains. In colder and wetter places, coarser sands give better aeration, but in hot localities or positions, they will dry out more rapidly. Advice on sand sizes and qualities may usually be obtained from builders' merchants and garden centers, which sell this material.

Gravels: These are small rounded or broken stones and for practical purposes include pebbles, crushed rocks of various types, crushed limestone and corals, silica gravel, river gravel, beach shingle, slate chippings, and haydite or burnt shale. The particle size of gravels generally ranges from one-sixteenth to half-an-inch in diameter. The best types have sizes of from one-eighth to three-eighths of an inch. For home hydroponics, using the simple method of growth, better results may be obtained by employing gravel aggregates that contain a reasonable proportion of finer material.

Broken Bricks: Bricks may be broken up into small pieces quite easily, using heavy hammers. It is best to reduce them to a maximum grade of half-an-inch sized particles, which will also give a substantial proportion of smaller chips and brick dust. Put all these, well mixed together, into the beds or containers to serve as the growing medium.

Vermiculite: This product is obtained from naturally occurring deposits located in various parts of the world. It is classified as a hydrated magnesium aluminum silicate.

The name comes from the Latin, *vermis*, a worm.

In the former, the scales are bonded together with water and in the latter with potassium. On heating to a temperature of about 2000° F, the water is converted into steam, which expands the material from twelve to fifteen times its original volume.

The resulting product is sterile, light in weight, highly absorbent, and retains water and air.

These properties make vermiculite very useful for hydroponic gardening. If you look closely at the small particles, you will see that they curl slightly and do look rather like short worms. Various grades of vermiculite are available in stores, garden centers, and horticultural suppliers.

It is best to ask for the standard garden grade. Vermiculite has given good results in soilless household gardens, but it is generally best to mix it with an equal amount of sand because its high water retaining properties may sometimes keep plants too damp during winter periods.

Cinders: Many types of cinders can be used for growing plants without soil. Soft and hard coal cinders should be soaked in water for twenty-four hours and then washed clean before placing in the hydroponic containers.

In old gardens, one can often find leach-troughs or tubs. The finer ash should not be discarded, but after drying, it should be mixed in with the bigger cinders to constitute a composite growing medium.

Other kinds of cinders include volcanic ash, which is in plentiful supply in countries like Grand Canary or near active or dormant volcanic zones. Charcoal is another type of cinder and makes an excellent hydroponic aggregate, though often rather costly and hard to get in industrialized areas. Lava can also be mentioned under this heading, and it can be broken up into a useful growth medium. Pumice is, of course, glassy lava, so full of gas cavities as to float in water sometimes.

Peat: This should not be utilized by itself as a hydroponic growing medium since its unbalanced composition may give rise to patches of poor plant development. However, small amounts of peat can usefully be added to sand and vermiculite aggregates, to loosen them up, as well as to improve aeration. The addition of between twenty-five and forty percent by volume of peat moss to coarse sand helps moisture retention.

Miscellaneous growing media: As well as the various materials already mentioned a few others had been used with success by soilless gardeners in different areas.

Leca: is a product of cement manufacture and is formed of numbers of hard, round pebbles with some dust-like residues. It absorbs up to four times its own volume of water. Readers may well find that in their district sources of alternative aggregates, not mentioned here, exist, quite suitable for hydroponics. It is always worthwhile trying such growing media, especially when they are readily available and cost little.

WATER SUPPLY

Most waters are quite suitable for home soilless gardening. Our supplies come from a vast number of sources, including rivers, reservoirs, wells, boreholes, and sometimes distilled seawater.

The first test is that if the water is suitable for drinking by human beings or animals, then it will be satisfactory for plants. At one time, it was thought that if the salt content of water exceeded two thousand five hundred parts per million, then it would be too saline for garden Flowers and vegetables.

Recent research work in North Africa and Israel has shown this to be incorrect. So long as there are good drainage and free movement of salty water through the growing medium, plants will tolerate very saline water, even seawater, for indefinite periods.

Many suppliers that would be unpalatable for drinking can do very well for raising flowers and vegetables.

Water containing a lot of magnesium and calcium salts is termed 'hard.' Both this kind of water and 'soft' waters are quite suitable for hydroponics.

Sometimes, waters do contain small amounts of salts that might injure plants. For example, excess chlorine will cause a hardening of growth.

To deal with this, it is best to filter the water through a tank packed with straw, dried leaves, or grass and then allow it to stand in the open for a few hours.

Rainwater is, in theory, pure, but in actual practice, it contains minute amounts of dissolved nitrogen, oxygen, and carbon dioxide, as well as certain impurities, especially near large industrial cities. These are not generally of any great significance in hydroponics. Municipal supplies, that is to say, ordinary tap water, have been filtered and treated. The chlorine content varies but is unlikely to be excessive for soilless culture.

For small household units, simple trials of water can be made by placing a few cut stems or flowers in a glassful over a period of a day or two. If no ill effects are observed, the water may be presumed satisfactory for use. These days it is quite simple to obtain analyses of most water supplies from the local municipality or city or in country areas from the agricultural departments. It is always exciting for the soilless gardener to know what type of water he or she is working with.

The application of water to hydroponic containers, or irrigation, as it is also called, can be carried out in simple soilless gardens by using cans, ordinary jugs, or hosepipes. In larger installations, it is done by means of systems of pipes and sprays, chiefly of course, in order to save labor costs.

SETTING UP A HYDROPONIC UNIT

Having selected the particular kind of container that you prefer, the first thing to do is to fill it with the chosen aggregate.

For beginners, there is little doubt that the most efficiently managed growing medium is a mixture of coarse sand with some more substantial material, such as broken bricks, gravel, cinders, vermiculite, or perhaps leca.

Alternatively, you can use freshly broken bricks or rocks, blending both the chips and the dust well together to form a composite aggregate.

If you decide on the sand and a separate second material, here is the procedure:

Take two parts sand and three parts of the larger aggregate, the grade of which should not exceed one Quarter-of-an-inch, measuring the amounts by volume, and mix them thoroughly. Do not weigh the quantities because the density is different, but use a tin or bucket to get the correct amounts.

Before you place this hydroponic aggregate in the container, spread about an inch or so of clean, broken flower pot pieces, small stones, or Pebbles on the bottom of the receptacle.

This will prevent any of the growing medium washing down into the drainage holes and perhaps blocking them up. It will also ensure better circulation of air. The rest of the container should then be filled with the prepared aggregate to within half-an-inch of the top. Firm it down gently and smooth it over, making sure that the surface is quite level.

As soon as the hydroponic container has been filled with the chosen aggregate mixture, water it carefully with enough clean water to moisten the growing medium. This will settle it in place. If you Have not done so already, put the unit in the position where you intend to operate it.

The watering may be done with a can, jug, or pipe, but it is much easier if you fit a rose or sprinkler to whatever appliance you use to spread the liquid evenly over the whole surface of the growing medium.

About half-an-hour after this watering, pull out the plugs from the drainage holes and allow any excess moisture to run off into saucers placed underneath receptacles like pots, or into trays and gutters for troughs, boxes, and beds. The plugs should be replaced when the seepage stops. The hydroponic unit will now be ready for sowing or planting.

In the following chapters, we shall discuss these tasks, as well as how to apply the vital nutrients and the general care of crops.

But before dealing with these essential operations, let us examine the question of suitable sites in or around the home and in the amateur's garden for soilless culture installations. One of the main advantages of hydroponics is that it can be started in places where ordinary soil growing would be impracticable. This opens the way to an extensive selection of positions for soilless units. Flowers and house plants will grow very well inside rooms, on window ledges, along with balconies, or in hanging containers.

Backyards, flat rooftops, the sides of paths and passageways are other apparent sites. For the kitchen, tiers of troughs for vegetables and quick maturing salads are quite practicable. Outside beds can be established on any open space, in Existing gardens, or on waste ground.

In cold regions, these may be covered with polythene sheeting on frames or cloches for protection, while in the tropics, matting shades will provide an adequate safeguard against the rays of the hot sun.

Roofing-felt is also satisfactory for the prevention of excessive surface temperatures causing scorch in hydroponic beds or troughs, and it can, incidentally, be employed as well to make proper containers if attached to wooden or brick supports.

The number of possible sites for soilless gardening is almost unlimited. A little ingenuity and thought on the part of the householder or amateur gardener will result in space-saving arrangements and adaptations of lasting value. So long as hydroponic units are kept in positions where the plants receive enough air, together with adequate light, natural or artificial, and a supply of water is available, the actual site itself is of secondary importance.

When you start soilless gardening, you will probably know just where you would like to put your containers. A little ingenuity will do the rest. Remember, however, that all sites must be clean and free from disease-carrying rubbish.

Dusty and windy places do not favor good growth, and plants will need protection by means of screens in such positions; otherwise, they may suffer severe damage. In houses, take care that gas leaks do not occur in the vicinity of hydroponic containers.

SUPPLEMENTARY TOOLS

In addition to the main items for preparing the simple growing unit, a few other things will be required for hydroponic home gardens.

Most of these may already be available in the house. For instance, a kitchen weighing scale or balance for measuring out nutrients; an old teaspoon; a hand fork for smoothing over the surface of the aggregate or moving seedlings and transplanting, as desired; a jug or preferably a small watering can fitted with a thin spout and rose or a hosepipe with a sprinkler attached for watering and applying the plant-feeding solution.

Further helpful accessories can include scissors or sharp knife for pruning and cutting, dark green twine and string for tying plants, bamboo canes for staking, a soft brush for dusting hairy-leaved ornamentals or pollinating cucumbers and tomatoes, and any similar useful equipment that may come in handy from time to time.

NUTRIENTS FOR THE PLANTS

Healthy plants demand ample supplies of nourishing foodstuffs and where these are deficient or unbalanced flowers and vegetables will grow slowly or not at all. The symptoms of food shortage are well known to scientists and experienced gardeners. It is the job of hydroponics to ensure that crops receive optimum quantities of plant nutrients, in the right proportions, throughout their useful lives. In the preceding chapters, we have learned the basic facts about soilless gardening, what the system sets out to achieve, and how green plants grow. In addition, we have also discussed the various kinds of containers suitable for home hydroponics and the aggregates or growing media that can be used to fill them and provide support for roots.

Now the time has come to talk about the mineral salts or fertilizers that supply the essential plant nutrients.

It has already been explained that when a correctly balanced chemical mixture or formula is dissolved in the proper amount of water, it forms a nutrient solution capable of sustaining crop growth.

Several hundred different formulas have been developed by different scientists and institutions concerned with soilless cultivation over the years, but they all have the same object: to supply plants with the vital food elements, such as nitrogen, potash, phosphorus, calcium, sulfur, magnesium, iron and other minor or trace nutrients. Today, it is generally accepted that the actual choice of mineral salts for fertilizer mixtures is of little significance, provided a balanced concentration of the necessary elements is assured.

Of course, care must be taken not to include incompatible chemicals in a formula or to choose those that might give rise to undesirable effects.

Given these guidelines, what then becomes of importance in practical soilless home gardening is the local availability of the nutrients and their cost.

Fertilizer salts usually are cheaper than laboratory grade nutrients. Therefore it pays to employ ordinary agricultural or horticultural brands.

Moreover, as will be explained below, since these commercial grades contain many of the minor elements such as manganese, boron, zinc, and copper as incidental impurities, it is scarcely necessary to add extra amounts of these particular nutrients to a simple hydroponic formula or mixture for household work, a fact that makes for easier maintenance and operation of units.

The amateur hydroponicist, starting soilless gardening, need only be concerned with one, or at the most, two, formulae for feeding his or her plants.

In fact, many firms do offer complete mixtures for use in hydroponics to the public. The question of whether to buy your nutrient mixtures ready-made or to blend your own at home is a matter of personal choice.

It is quite simple to weigh out fertilizer salts in the kitchen on an ordinary balance or scales and then mix them together. It is also cheaper in the long run and adds greater interest to your hydroponic gardening.

The fertilizer salts mentioned in the formulae recommended here should be readily available in all parts of the world, especially from agricultural and horticultural firms.

NUTRIENT FORMULAE

For general use in home soilless gardens, the following mixture will give excellent results.:

MIXTURE 1:

- Quantities are in grams:
- Sodium nitrate 12
- Potassium sulfate 355
- Superphosphate <3
- Magnesium sulfate142
- Iron sulfate 100

Enough to cover the head of a match:

- Nitrogen
- Potassium
- Sulphur Phosphorus
- Calcium Magnesium
- Sulphur Iron

MIXTURE2:

- Ammonium sulfate 10

- Potassium nitrate 9

- Monocalcium phosphate 4

- Magnesium sulfate 43

- Calcium sulfate 3

- Iron sulfate 6

Enough to cover the nail of your little finger:

- Nitrogen

- Sulphur Nitrogen

- Potassium Phosphorus

- Calcium Magnesium

- Sulfur Calcium

- Sulphur Iron

Note: In the above mixtures, trace nutrients, including manganese, boron, zinc, and copper, will be present as impurities in the fertilizers listed and probably in the water supply.

You will notice that the quantities of the individual fertilizer salts vary in each mixture.

This is mainly because different salts contain different percentage proportions of the range of nutrient elements. The fertilizers are, in fact, merely the vehicles through which the elements are made available to the plants' roots. Hydroponic formulae are compiled according to standard calculations, which will be found in more advanced books or works on agricultural chemistry.

PREPARING MIXTURES

The prescriptions listed above should be prepared in the following manner:

Carefully weigh out on a kitchen balance the lots of the individual salts that you will see in the formula you have selected. As you do this, put them in turn in a bowl or other container, and when you have finished weighing, mix them all well together using a wooden spoon, pestle, or other blunt instruments to break down any lumps.

The resulting mixture should resemble a reasonably fine powder. Sometimes it is best to crush the iron sulfate separately and add it last to the mixture to ensure better distribution of this small amount of salt.

The formulae must always be stored in a dry container, with the lid closed. Never let nutrients get damp or wet before application.

This also refers to the special fertilizers, which should be kept in a suitable store place.

To prepare more substantial quantities of nutrient formula, simply multiply up all the ingredients to the desired amounts, using a constant number, so that the relative proportions remain the same. It is important to remember this; otherwise, any alteration of the inbuilt ratios will make the mixture unbalanced.

SOWING IN HYDROPONICS

Once the hydroponic containers have been filled with aggregate and given their first watering with plain water, sowing or planting may commence. The growing medium should have been well dampened, being just as moist as a wet sponge that has been lightly wrung out. The drainage holes will be closed with the plugs that you have provided, after the initial aeration of the aggregate.

Good sowing helps plants considerably in their early lives and indeed throughout flowering and fruiting. In hydroponics, the task of seeding is made much more accessible than in soil gardening because the aggregate is of generally even texture throughout and at the right degree of moisture, as well as being free from contamination. Moreover, only a light covering of a growing medium is necessary to conceal the germinating seeds from the light.

Avoid overwatering newly-sown seeds; the aggregate around them should be as moist as a lightly wrung-out sponge. This permits enough moisture to reach the germinating seeds while allowing essential aeration. Usually, large seeds are put into small holes made with the end of a pencil, or something like the handle of a spoon, in the aggregate, about half-an-inch deep, at intervals, so that when the growing medium is smoothed back over them, they will have a covering of approximately half that amount. It is not necessary to bury seeds deeply in hydroponic gardening. When sowing smaller ones, the easiest way is to remove a little aggregate from the surface, sprinkle the seeds carefully, and then replace the growing medium on top of them, to give a thin but effective covering — mixing excellent seeds with some dry sand or ' filler' assists in securing even distribution. Pelleted seeds may now be bought in many places and are extremely simple to sow. They can be inserted into the aggregate at appropriate intervals.

Usually, the maximum depth at which most flower and vegetable seeds should be planted should be such as not to have more than a quarter of an inch of aggregate over them. Seeds need to germinate in the dark under a protective, but light, the cover of the thin layer of surface growing medium. In hydroponics, it is possible to plant much closer than in soil, since ample nutrients are available. However, the number of seeds placed in any single soilless garden container will depend upon individual preferences and the need to allow sufficient light to reach each plant. Therefore, follow standard distances in general, remembering that you can reduce the spaces between plants by up to fifty percent, if the circumstances warrant this, under favorable conditions, and you want a closer or thicker stand or display.

Let us take for an example the sowing of a hydroponic trough or container about two feet long by one-and-a-half feet wide, set up by the kitchen window.

To begin with, you want to try easy-to-grow crops like lettuces or dwarf tomatoes.

Take twelve good seeds of the plants you prefer and sow them by pushing them gently down into the moist aggregate with a small stick.

Do not bury them more than half-an-inch at the most. A hydroponic trough of this size will have enough space for at least six plants. So put the seeds two together, at equal intervals, with three groups of two each down one side of the container and the same number along the other side. See that they are covered nicely by the aggregate and not left exposed to the light. When the seeds have germinated and are growing well, pull out the weaker one in each group and discard it, to permit the healthy plant to develop by itself.

The object Of sowing two seeds together is because sometimes a seed may fail to grow well, and it saves time to have a second one ready to take its place.

When sowing has been completed, smooth over the surface of the aggregate in the containers, making sure that no seeds are exposed. If one or two are, push them down gently into the growing medium so that they are appropriately covered. It is a good plan to give a very light watering soon after sowing. This can be done using the technique already recommended a watering can be fitted with a rose or sprinkler, a jug with a sieve to pour the water through gently or even an old teapot.

Rubber sprinklers are readily obtainable from garden shops and can be slipped over the spout. After a few days or a week or two, depending on the kind of seeds sown, germination will be noticed. Tiny green plumules or shoots will appear above the surface of the growing medium. If you have sown too many seeds in a container, remove or thin them out or transplant them to other receptacles when they are about three-quarters to one inch high. A kitchen or small hand fork is excellent for this job. Also, it is easy to rearrange the tiny plants in different positions at this stage, if desired, in the dibber or stick for making holes in the aggregate

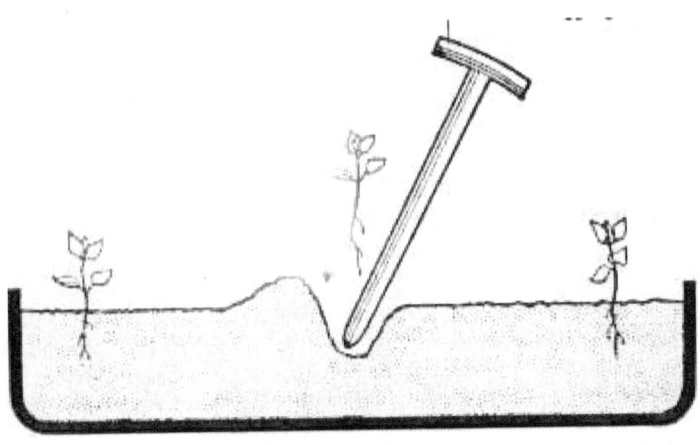

Transplanting a seedling.

After making a small hole in the aggregate, drop it in carefully with roots nicely extended, not bunched up, then gently push back the growing medium around it so that the young plant stands firmly.

Always be gentle and careful not to damage the tender seedlings; harsh pulling breaks the roots. Properly moved, there will be no check at all to hydroponic plants.

If you wish to use young seedlings, probably bought from a garden shop or nursery, instead of seeding your containers, be sure to wash off any soil adhering to their roots before planting them in a hydroponic trough. This can be done by holding them under a slowly running tap. Make small holes in the aggregate at equal intervals, as for seed sowing, drop in the seedlings, taking care that their roots are not turned up and gently rake back the growing medium around them. The stems should be firm and steady and not left to wobble about. Many hydroponic gardeners like to raise their own seedlings in separate boxes of sand and then transplant them into the main containers later.

SOLUTION APPLICATION

Assuming that you have seeded or planted your hydroponic pots or troughs and given them a very light sprinkling of water to smooth down the surface, the task now is to ensure that the growing medium is kept continuously moist with nutrient solution containing the vital plant foodstuffs. We have already seen how the fertilizing formula is prepared, and if you have not already chosen the one you mean to use and mixed it up or bought a proprietary formula, you should do this immediately.

Take the correct quantity of nutrient mixture that is one-third of an ounce (ten grams) or a level teaspoonful and dissolve it in a gallon of water. Stir it well so that no residues remain at the bottom of the can, jug, or other vessels.

If you think, or know, that you will require more than a gallon of solution, mix up more significant amounts by increasing the weight or number of teaspoonfuls of formula and the quantity of water.

Naturally, bigger cans or vessels will be wanted for higher quantities of nutrient solution. In large hydroponic units, tanks are employed for mixing up the solution. For proprietary mixtures, follow the manufacturer's instructions exactly.

The day after sowing or planting the hydroponic containers, or even the same day if the weather or site is warm, apply the first dose of the nutrient solution by watering it well over the surface of the aggregate.

Sprinkle it evenly so that all parts of the trough or pot get fair amounts. You can, of course, use a hosepipe fitted with a

Spray or rose in bigger garden units to save time and labor. The aim is to maintain the growing medium in a continually damp condition, resembling a wet sponge that has been gently wrung out.

The importance of proper solution application or irrigation cannot be over-emphasized. Make further applications at necessary intervals so that the growing medium does not dry out at any time. At no period, either, should it become waterlogged.

That is, there must be no excess of surplus solution standing on the surface. Every few days, Remove the plugs in the drainage holes or the Plasticine seals for a short while to allow any remaining water or solution inside the bottom of the containers to seep out into the saucers or trays and gutters provided to catch it.

This also makes it possible for extra air to move through the aggregate from the open holes to the surface of the medium, for the benefit of the roots. Usually, an hour or two's aeration will suffice at a time.

Sometimes you may find, when the weather is warm, or the plants are growing in a centrally heated room, that daily applications of the nutrient solution are desirable, but in general, two, three, or four times weekly may be enough.

Do not delay the application of the plant food; it should commence not later than the day after sowing or planting and continue regularly, as often as needed, right through the lifetime of the crops. The simplified method of home hydroponics described in this book has been designed to ensure that plants are neither overfed nor underfed, as they often may be in conventional soil gardening.

Instead, a fixed and balanced amount of nutrients is given to provide for extraordinary and rapid growth. Naturally, quite a lot depends also on the preparation of the containers, the necessary drainage holes and their usage, and the correct filling with aggregate.

The operations of watering and feeding plants have been combined into a single task by applying the nutrients in solution. Provided the instructions are adhered to, plants will not be damaged or starved at any time. The quantity of solution that you prepare at intervals will, of course, depend on the size of your soilless garden.

As growth proceeds and the seedlings get taller and change into big plants, the frequency of solution applications may be increased. Take care always to employ a sprinkler or rose on the watering-can, jug or hosepipe, so that the surface of the aggregate is not unduly disturbed. In summer months, especially in gardens outside or in warm houses, the application of plant food may be carried out daily or every other day, although conditions vary, and it is not possible to stipulate the exact times.

During the winter or cold periods, longer intervals between applications will be practicable, because evaporation will be less.

The golden rule is to see that the aggregate in the hydroponic containers is always as moist as a damp sponge that has been squeezed out lightly.

This bears repeating and must be remembered if good results are to be obtained. If it is kept like this, first-class growth and healthy development will follow. In using the nutrient solution, you are providing both water and foodstuff together, a process that saves both work and time.

Remember, never overwater with the nutrient solution; this prevents air from moving through the root zone and may cause wilting and death of the plants. Watch for any excess moisture and open the drainage holes to let it seep out. Once every week or ten days, it is advisable to flush through the aggregate gently by pouring some plain water (without added nutrient fertilizers) onto the surface through a sprinkler or sprayer. This will flow down, carrying away any accumulation of unused plant food, and it should be permitted to run off through the drainage apertures and then discarded.

Don't forget to replace the coverings or plugs in the holes before making the next application of a fresh solution. For house and conservatory plants, try to use water and solution at room or greenhouse temperature. Tough waters can be treated with a little commercial water softener. Rainwater is ideal, but very heavily chlorinated water is best left to stand in a bucket for a day before application. Plants will benefit from periodic spraying of the leaves with clean water to wash off any dust. During the resting seasons of foliage, types reduce watering or irrigation with a solution considerably. Another point to note is that if you let a solution or water stand or lie too long on leaves when they are exposed to strong sunshine, severe scorching may occur. In the tropics, overhead shades in hydroponic units will stop this sort of damage to crops. Morning watering is preferable in unheated rooms in winter months.

LIGHT

Light is vital to plant growth, but different species vary in their demands for illumination, as every gardener knows. It is advisable to keep your hydroponic units in the best light you can, preferably near windows in the case of indoor installations, especially in wintertime.

White or cream-colored walls and ceilings give a better light reflection in poorly-lit rooms.

Leaves and stems will bend towards the source of light; a phenomenon called phototropism causing plants to grow in a lopsided manner in certain situations, so pots and other containers may have to be turned around at intervals. Young seedlings should not be exposed to bright sun for very long periods. Where sites are too dark, it is quite simple to install forty-watt 'daylight' type fluorescent tube lighting, with reflectors, to supplement the ordinary electric bulbs.

These tubes do not emit as much heat as filament lamps and so will not cause leaf burn. They also arc pleasanter for general home use. Plants can be kept within a distance of a foot or two of fluorescent lighting, but should not be near ordinary bulbs.

AIR

Together with air, we also have to consider humidity. Ventilation in closed hydroponic units or household installations is essential. There must be ample circulation of fresh air, but draughts, gases, dust, and smoke arc harmful to plants. However, modern gas appliances, properly fitted, are stated to be quite safe and should not emit any noxious fumes. Fires and radiators dry the atmosphere so that it becomes necessary to raise the humidity in order to keep conditions healthy.

This can be done by having a pail or pan of water standing in an inconspicuous place in a room, by spraying leaves with a syringe or sometimes wiping them carefully with a damp sponge or moist cloth or by using an electric humidifier. Steam from kitchens and bathrooms also helps, quite often, to improve the humidity in houses.

It is worth remembering that too low an air moisture causes the death of many household plants and greenhouse flowers and fruits. The amount of desirable moisture in the atmosphere is assessed by categories of relative humidity. These are:

high, over 70%;

moderate, 50% to 70%,

low, under 50%.

Excess humidity encourages the spread of disease.

TEMPERATURE

Generally, most plants prefer an even degree of warmth, without significant fluctuations. Provided the range is between 50 ° and 75 ° F, the vast majority will do well. In rooms and houses in cool places, the amounts of light and moisture in the air are not excessive, and the same thing applies in hot, dry or arid districts.

Too high temperatures with weak light and considerable dryness usually result in poor growth and shriveled-up foliage.

Up to an optimum, plant growth may be improved by increases in warmth. A fall in temperature usually reduces development, while at or below freezing point, the death of many plants occurs.

Most species have been accustomed for countless generations to certain limits of heat and cold, and though it is possible to acclimatize plants in new areas, the process of so doing is often wearisome and slow.

The temperature of plants tends to follow that of their surroundings, but it may be higher or lower, owing to the fact that vegetation responds more gradually to heat changes than does air. Daily alterations are commonplace, while wind, cloud, and seasonal differences exert significant influences. For most temperate types of plants, the best heat range is from 60 ° to 70 ° F, and from 75 ° to 90 ° F for tropical species. A maximum of as much as 125° or 130° F can be tolerated by torrid kinds; indeed, growth is known to proceed at even higher temperatures than these.

Minimum temperatures vary from species to species. Some plants will be killed by frost, but others survive all year round.

Do not put household plants near hot radiators at night.

If you think it may freeze inside rooms, in greenhouses or under cloches, then you can cover them with newspapers or cloth, and draw curtains or screens across glass or window panes in wintertime. Frost is a great killer, but sun scorch and desiccation are equally fatal in hot countries, where overhead shading is often essential, or outside plant houses may have to be erected in the garden or compound. Growth is generally far slower at cold periods or in times of drought when the air moisture is extremely low, but it is often surprising how tolerant some plants may be of adverse conditions and how they adapt to new situations.

In horticultural terms, temperature ranges are graded as follows:

hot, above 80° F;

warm, 65 0 to 80° F;

moderate, 50 0 to 65 0 F;

cold, under 50 0 F.

At 32 F, frost will occur, but this is unlikely inside well-built houses.

Nevertheless, when room warmth or greenhouse temperatures fall as low as 45° to 35° F, the conditions become difficult for most plants, and indeed for human beings too.

CLEANING

Dust can be a problem in household hydroponic units.

It prevents plants from breathing properly by blocking up the pores in the leaves. It can be kept in check by syringing periodically with clean water.

A soft cloth or light brush is useful for ornamental and foliage types. Always observe strict cleanliness in hydroponic home gardens, removing dead leaves, stems, rotten fruits, or finished blossoms at once and wiping pots and containers regularly. Rake over the aggregate from time to time lightly with a small hand fork or rake. Hygiene will pay good dividends.

DAILY WORK

The general care of simple hydroponic gardens should present no problems for beginners.

There is no hard manual labor, no digging, weeding, or similar tasks.

The first job is to keep the soilless garden free from dirt and rubbish. Dirty conditions cause disease, and insect pests will soon make an appearance if vegetable refuse is left lying around. In uncared-for greenhouses, a red spider is likely to become a nuisance.

The primary routine duty for hydroponics is to check the condition of the aggregate regularly in the containers and see that it is appropriately moist and the plants are thriving.

It is worth remembering that all living organisms are subject to change, so no matter how good a technical system of cultivation may be, the condition of the plants can alter from day to day.

There is no substitute for careful observation. Once you get the 'feel' of hydroponics, which you will undoubtedly acquire after a few weeks of operating a soilless garden, you will be able to sense how the crops are doing and tell at once the responses of the plants to the existing environmental and nutritional conditions.

This feeling for hydroponic plants may well be the equivalent of what is popularly known as 'green fingers' in soil gardening. It is difficult to define, but it is none-the-less authentic.

No doubt this is why soilless cultivation is an art as well as a science. The more sensitive the grower, the better will be the results.

Day by day observations of nutrient solution needs, which will vary according to the time of the year, the particular plants growing, and the location of the hydroponic garden is vital to proper management. Such factors as air moisture, the strength of wind, temperature range, and ventilation may be very significant. The aggregate or growing medium should be at the right degree of moistness. Excessive wetness stops proper aeration, while if the root crowns or tops are waterlogged, the crops will soon die. Improper aeration quickly becomes apparent in hydroponic troughs or pots.

The plants assume a tired look—that of age prematurely imposed upon youth. It is easily detected after a little experience. Regular checking of light, warmth, and other changes is most beneficial. In very hot, dry weather, shades should be spread over hydroponic gardens, particularly outside ones.

When heavy rain occurs, protection is desirable through some form of the canopy to stop excessive flooding of the containers.

Moderate rain or showers will irrigate beds or troughs quite satisfactorily and offer no problems, so long as they do not dilute the nutrient solution present in the growing medium too much. You will soon see if this has happened after a few days when growth slows up, or the crops look a bit pale in color. To adjust the deficiency, simply put screens over the unit and apply other solutions, excluding rain for a period until the plants look better. In greenhouses or where plants are sited in picture-windows, attention will need to be paid to the light conditions, opening or closing shades as required. As the time of harvesting vegetables or picking flowers and fruits approaches, growers should make frequent inspections of the condition of the plants and decide when they are ripe or ready for collection. It is both exciting and helpful to keep records and notes about your hydroponic garden.

Nothing is too insignificant to enter on such charts. A well-kept notebook becomes in time almost a text in itself and constitutes a valuable guide for future work.

The dates of sowing or planting can be written on labels attached to each container. Inside houses or in greenhouses, plants like tomatoes and cucumbers will need to be pollinated to ensure proper fruiting.

This may be done by gentle spraying with clean water at a warm air temperature or by rubbing the flowers very gently and carefully with a piece of soft cotton wool. Transplanting in hydroponics is simple. The seedlings or young plants should be removed gently with a small fork from the aggregate and inserted in holes made with a pencil end, a piece of stick, or a thin dibber, in their new situations.

Smooth back the growing medium around the roots, so that they stand firmly. Apply a little solution at once, and there will be no check-in growth. Repotting can be done by up-ending the container, holding the plant and aggregate around its roots securely with the hand, and then placing the whole mass in the new pot or other receptacles.

If these containers are larger than the old ones, add a little extra aggregate to fill them appropriately.

Repotting or replanting may have to be done if the roots of perennials grow down into the drainage holes, or if the plant is obviously too large for its trough.

Pruning and training may follow standard lines. The general idea is to improve growth habits, prevent plants from straggling or getting out of hand, and from preventing new branching.

Aways cut out dead or diseased stems as soon as they are seen. Supports may be employed where stems are brittle, or flower heads and fruit bunches or trusses are very heavy, as well as for climbers.

Normal methods of staking and tying can be used. The best way of supporting hydroponic plants is to employ strings or twine attached to overhead wires. This avoids inserting canes or sticks into the growing medium. In planting potatoes or bulbs, slightly larger holes will have to be made in the growing medium to accommodate the tubers or corms. All root crops do well in hydroponics.

So does asparagus, fruits such as strawberries, vegetables like celery and chicory, and indeed any other kind of garden produce. Mushrooms are not, of course, higher plants but fungi, and special soilless composts are needed for their cultivation.

After plants have finished flowering or fruiting and are not of any further use, remove them from the containers, shake the aggregate off their roots, and discard them. The growing medium can then be flushed well through with plain water, smoothed over, and utilized immediately for new sowings or plantings.

Don't waste any aggregate; it is made up of inert materials and, when cleansed of any accumulated mineral salts and root debris, will continue to give excellent service for years.

There is no rotation of crops in hydroponics, and the same growing media may be employed over and over again without risk as long as these simple rules are followed.

Provided enough warmth and light can be supplied from whatever sources, you can grow flowers and vegetables in the home or garden all the year round in hydroponics, irrespective of the season.

This great advantage of soilless gardening, making possible the culture of plants out-of-season, Is significant and constitutes a real boon for householders and amateurs.

Not only can it bring considerable pleasure and enjoyment, but it also may help you to save money by growing your own produce in the home, thus avoiding the expense of buying costly out-of-season greenstuff for the family.

GROWING HEALTHY PLANTS

Hydroponic home garden units seldom give trouble when operated according to the simple method recommended, and your soilless grown plants will remain healthy if you follow the easy rules laid down and take standard precautions. Try to satisfy the needs of each lot of plants. Spend a few minutes daily looking them over, watch the leaves, stems, flowers, and fruits for any warning signs, and remember that love and care are vital. Hydroponics is a scientific and controlled system of growing, but, as in all such activities, when we are dealing with living organisms, a little human interest will be well rewarded.

IN CASE OF TROUBLES

The first thing to do is to prepare a checklist covering such items as light, air, moisture in the aggregate, fumes, situation, hygiene, and other significant points. Go through this, noting any likely causes. Light deficiency can give rise to poor blooms, weak stems, often elongated, and very pale leaves, a condition known as chlorosis.

Sudden collapse may be caused by noxious gases or smoke and fumes. The leaves are excellent indicators of a plant's state of health. Leaf drop can be started by several things:

overwatering, too dry an aggregate, or very sudden changes in temperature are some of the reasons for this trouble. Yellow or drooping leaves are frequently traceable to an excessively wet growing medium, with the drainage holes not being opened regularly or failure to feed correctly with nutrient solution.

Black leaves are dead, and frost and sunscald may be the culprits, while browning of the tips often arises from overfeeding (that is, not using the correct amounts of the nutrient formula in the water), exposure to draughts, and possibly too much liquid standing around the crown of the plant where roots and stem join. Leaf spotting is commonly ascribed to sunscald or watering from above in strong sunshine.

Rotted stems come from too wet roots and over-application of solution. Stunted growth with small leaves, a halt in development, and lack of bloom on flowering plants can be attributed to inadequate nutrition and light or failure to look after the particular needs of different types of plants.

A common cause of trouble with household plants is the lack of humidity in the atmosphere.

Do not try to treat complaints by applying all suggested cures at once; use your checklist to eliminate the various possibilities in turn and endeavor to test each remedy in succession until you find the exact problem and its answer.

DISEASES AND PESTS

Neglected plants frequently fall victim to diseases and pests far more quickly than do those that are vigorous and well cared for. Provided the simple instructions given in this book are followed, and the hydroponic unit set up correctly, soilless gardening for the home is remarkably free from the diseases that beset soil cultivation. Remember, though, not to import disease organisms by sowing contaminated or doubtful seeds in your containers or by buying seedlings grown in soil from nurseries and shops and planting them, unwashed, in your soilless unit. That is asking for trouble. Diseases such as rot and mildew or other fungus infections are generally encouraged by overwatering or using excess quantities of the solution and by too damp conditions, lack of ventilation, and overcrowding. Good hygiene and proper regulation of the environment are the best preventives.

Various fungicides are available for spraying plants that are attacked.

The most widespread pests include ants, aphids or greenflies, caterpillars, mealybugs, red spider mites, scale insects, thrips, and whitefly.

A number of insecticides recommended for destroying these nuisances are readily available from garden centers, chemists' shops, and other stores. Liquid derris and pyrethrum-based sprays are useful and quite safe to use in the home or garden but always cover fish tanks or bowls in the vicinity before applying insecticides.

For scale insects and mealybugs, however, treat by rubbing them off gently with a matchstick tipped with cotton wool soaked in methylated spirits, if time permits. Ant repellents may be bought easily.

Usually, in clean surroundings, the incidence of insect pests will be very much less than in places where rubbish is allowed to accumulate or where dirty conditions prevail.

Sometimes, a slight greening of the surface of the aggregate in the hydroponic containers may occur, indicating the presence of algae. To check this, water occasionally with a solution made up of 0-5 of a gram of copper sulfate dissolved in two gallons of water, or pro-rata.

If the trouble is extensively repeated, waterings may be given every two or three days until the algae disappear. Copper sulfate may be obtained and weighed out at any good chemist. Algae infection typically arises mainly when conditions are somewhat damp.

AVOIDING ERRORS

Prevention is always better than cure. Experience over many years shows that failures in hydroponics occur most frequently because the growing instructions have not been followed. So please bear in mind that you are operating a scientific and controlled method of plant culture. Simple indeed, it is, and perhaps because of this, you may be tempted to forget the few straightforward rules occasionally. For plants in household and amateur hydroponic gardens, the main points never to be forgotten are:

- Always see that the aggregate in the containers is as moist as a damp sponge that has been wrung out lightly. Do not over-irrigate the growing medium or let it get too dry, and open the drainage holes periodically to give aeration.

- Give the particular plants you are growing the best environment and conditions that you can, so that their needs for light, ventilation, warmth, and other physical requirements are met as far as may be practicable.

- Follow the mixing up instructions for the nutrient formula you choose and for the preparation of the solution of plant food most carefully. Be sure to use the right quantities of the fertilizer mixture to the stated amounts of water. Don't guess at these and remember to apply the solution regularly, as the season demands, according to the crops' needs, to keep the aggregate moist and the plants vigorous and healthy. There is nothing complicated about this, and a few days' experiences will show you just how much solution will bring the growing medium to an excellent degree of moisture and maintain it there at different times.

- Maintain strict hygiene: in technical terms, we call these proper phytosanitary precautions. Be a keen observer and get to know each group of plants. Your care and patient attention will be well repaid.

FLOWERS

Soilless gardening is a hobby that combines both science and art, affording ample scope for ingenious and attractive arrangements and floral displays. With hydroponics, you can be sure of reasonable technical control over plants. Beginners will find home growing without soil can bring much satisfaction and many exciting and happy hours spent in the operation of hydroponic units. All the same, home gardeners succeed best when they develop an awareness of the needs and habits of different kinds of plants. There is no real substitute for abundant and loving care and a sharp mind.

Today, most householders, given space, will also be amateur gardeners. The value of soilless culture is that, even when conditions are congested, the chance of having a garden, small though it may be, is not lost. New fashions in flowers have appeared, with varieties for contrasting requirements. Year by year, the popularity of homegrown blooms increases.

The urge to have flowers and foliage in and around homes is ancient.

The Romans grew plants in pots in their villas, and, for centuries, both the Japanese and the Chinese have specialized in miniature indoor gardens, as well as larger outside ones.

In Europe, covered orangeries were once a feature of stately houses, while cottages and town residence windows used to be bright with fuschias and geraniums in the summertime.

Many tropical lands are well known for the beautiful plant houses attached to the homes of famous people.

Still, many amateurs and housewives continue to suffer grievous disappointments in their efforts to grow garden flowers and indoor plants. Millions of plants die quickly owing to the failure of their owners to look after them properly or because of wrong cultural advice.

Then, too, most specimens bought from stores, nurseries, florists, and markets have been forced into unnatural development by artificial means before the sale, or maybe infected with diseases not apparent to the untrained eye at the time of purchase. The soils or composts in which they are rooted are frequently unsuitable or rapidly become exhausted. Consequently, such plants soon languish and die.

It is partly to avoid such misfortunes and losses that the new easy-to-use hydroponic method discussed in this book has been devised, while the simple growing techniques will provide a reliable means of starting the beginner on the right lines.

CHOOSING FLOWERS

The first thing to do is to decide precisely why you want each lot of plants. Very many types of house and garden flowers have been grown successfully in hydroponics.

Perhaps the highest amount of work has been carried out with carnations and chrysanthemums.

While much depends on local conditions, including things like temperature, light, and humidity, there is usually no problem in raising all kinds of flowers in properly run soilless gardens. Flowers are usually grouped into the following categories:

- Alpines are, Strictly speaking, plants from mountainous regions, but loosely used to refer to all dwarf types suitable for rock gardens. They may be hardy herbaceous perennials, biennials, annuals, or shrubs.

- Annuals (Half-hardy) These complete their cycle of growth in one year but need protection from cold in certain regions during springtime.

- Annuals (Hardy) These are resistant to cold.

- Aquatics Plants that grow wholly or mainly in water.

- Bedding plants This is a garden term only, having no botanical significance. It refers to those ornamentals that can be used for massed effects.

- Biennials Plants which complete their cycle of life in two years. They do not usually flower in the first year.

- Bulbs There are hardy and half-hardy bulbs as well as tender kinds.

- Herbaceous perennials Plants that live for a number of years and have a comparatively soft growth. They may be hardy or tender.

- Epiphytes Plants that may grow on other plants without being parasitic.

- Shrubs and climbers

- Roses

- Miscellaneous types Including flowering and decorative herbs for scent.

THE ART OF ARRANGEMENT

Plants bring into our homes a breath of life and exceptional natural beauty. During dull winter months, their colors cheer us up, while on hot summer days, the green foliage and delicate blossoms look crisp and refreshing.

With a little imagination, very charming and tasteful interior decorations may be achieved with floral displays. A wide variety of arrangements for growing hydroponic flowers indoors, in backyards, greenhouses, or in the garden area, as well as house plants, are easy to devise with little trouble or effort. Here are some exciting ideas for soilless home gardeners.

Troughs and pots: Ready-made troughs and pots of different lengths or sizes can be bought at garden shops, department stores, or other suppliers. It is best to choose stone, pottery, or plastic types, although wooden boxes and containers are suitable if you line them with polythene sheeting. If metal kinds are used, be sure that they are painted inside with non-toxic paint, emulsion, or varnish.

This is particularly important in the case of galvanized iron troughs and receptacles.

Beautiful specimens are termed jardinieres. For most plants grown in home hydroponic units or soilless gardens, troughs and pots should be about six inches deep to allow ample space for rooting. To make pots more attractive, you can get hiders of fancy wire, cane, pottery, or colored plastic, or you can even use copper bowls and small buckets. Single pots standing by themselves should have saucers placed underneath to catch seepage when the drainage holes are opened.

In the case of troughs, one or two tin lids or a shallow tray will do the same job. Often containers of pleasing designs can be bought with attached legs, and others can be set up on stands to suit individual wishes. It is also effortless to make your own hydroponic receptacles at home, using wooden boxes, roofing felt, or polythene sheeting stretched over frames, or asbestos, while in the greenhouse or outside garden, beds, and troughs of bricks and several alternative methods of using window boxes or troughs are practicable.

In summer, the containers can be put out on the sills. At other seasons, it is best to have them on ledges inside, where adequate light is available, while the glass panes afford protection from cold and frost. Especially delightful are hydroponic plant windows, with long troughs running across the full width of picture-windows.

These heighten the effect of the area of glass and break up its monotony.

Venetian blinds can be employed to exclude sun scorch or excessive cold. Climbing displays are also useful for framing windows.

Tiers: All sorts of tiered hydroponic arrangements can add enjoyment to home hydroponic gardening. For example, shelving or staging, with various containers distributed at appropriate intervals, will provide screens, room dividers, door frames, and similar decors. There must be enough space left between tiers to allow plants to grow to their desirable heights.

Wall displays: Special wall-mounted containers may be obtained, and these usually include a drainage saucer. These look good in halls, corridors, or along staircases, as well as in rooms or beside windows.

Terror turns: These are smaller self-contained groupings, enclosed in carboys, large bottles, globes, fish bowls or tanks, and even big drinking glasses.

Generally, bowls are employed, with reproductions of Lilliputian or small-scale garden models, including suitable plants and toy summerhouses, pagodas, bridges, or windmills, together with paths and related embellishments. Somewhat similar are indoor groups of plants growing in a single container, combining tall and shorter specimens.

Other suggestions: Some people are very fond of bark and log arrangements consisting of attractively positioned stumps or cut branches and suitable plants. The pot-et-fleur containers are mixed indoor groups, with tubes inserted at intervals in the growing medium to hold cut flowers in place, which can be renewed or changed from time to rime to provide variety and extra brilliance. Hanging baskets arc normally made from wire and are lined with moss, in which a bowl or pot is put to act as the plant growth container. For climbing plants, pendant strings or chains or light canes will give support, and it is also possible to set up a trellis. Herbs do well in pans, especially being suited to the kitchen or pantry, while fruit pips and stones placed in pots will develop into tiny trees of fascinating appearance.

The cut-off tops of carrots, turnips, beetroots, and parsnips, planted in shallow containers, will soon give rise to fern-like masses of feathery fronds, useful for background greenery. Pineapple tops may similarly be cultured without trouble. Indeed, the enterprising amateur gardener or householder should have no difficulty in thinking up all sorts of exciting and conventional floral and foliage arrangements for indoor hydroponics.

VEGETABLES AND SALADS

The word vegetable is derived from the Latin term *vegetabilis,* meaning *animating.* Fresh green food is an essential constituent of all balanced diets.

The importance of vegetables and fruits for the maintenance of good health and bodily vigor is generally recognized today. During recent years, the popularity of the salad vegetables has increased steadily, so that now there will be few households in which tomatoes, lettuces, cucumbers or other fresh produce of similar type do not find a place.

Correspondingly, the interest in the basic greenstuff's, such as cabbages, spinach, brussels sprouts, green beans, and cauliflowers, as well as in aubergines, sweet peppers, and other more exotic produce, has also risen, since, once accustomed to the taste of fresh vegetables, people are loath to do without them in some form or other at times when perhaps their first choices may be unavailable.

In-home hydroponic gardens, you can grow all the vegetables and salads you may desire, together with some fruits and root crops or potatoes, if you wish.

The exact selection of the plants will, of course, depend upon the personal preferences of the housewife or amateur gardener and the space available. Local climatic conditions and facilities for giving protection during periods of colder weather are further factors that have to be taken into account. But it is worth remembering that hydroponic vegetables and fruits mature more quickly than those planted in soil, produce far higher yields, and need less space.

RECOMMENDED CROPS: Tomatoes, green beans, cucumber, aubergine or egg-plant, sweet peppers, radish, green or spring onions, lettuces, turnip, greens, mustard greens, Chinese cabbage greens, Swiss chard, celery, and spinach.

Of course, we can grow many more, and this list only mentions a few types that are extremely simple to cultivate hydroponically. Before going on to discuss these kinds of vegetables, and others as well, in further detail, there are three points about cropping in soilless gardens that are worth remembering.

First of all, you can take advantage of what is called intercropping. Some plants take up more space with their foliage or leaves than with their roots, thus leaving a free area of aggregate at surface level.

Between such tall plants, lines of smaller specimens that appreciate a little shade may be sown or intercropped. Examples are lettuces between rows of tomato plants, radishes among peas, and endive interspersed with French or runner beans.

Then there is catch cropping. This means that while larger plants are developing, seeds of quicker growing catch crops may be sown in the troughs or containers. These latter will be ready for harvesting before the former need the extra room. Examples include salad greens planted among beetroots, dwarf beans with cabbages or mustard, and cress with celery.

Finally, crops of different habits can be grown together in the same trough or bed with success, so saving an appreciable amount of space. Pumpkins or vegetable marrows and maize (sweet corn), tomatoes and new potatoes, or climbing beans and various root vegetables such as carrots, lend themselves to this combination.

Here is a list of the main vegetable crops, with some notes on looking after them:

Artichokes (Globe Or French): This is a perennial vegetable, which continues to reproduce for several years without replanting. Artichokes like good moisture but will not tolerate extremes of temperature. Protection from frost is necessary and shading in very hot weather. Cropping is most profitable in the second and third years.

Artichokes (Jerusalem): This does well in hydroponics. As growth proceeds, a little extra aggregate should be drawn up towards the stems to make sure the tubers are properly covered. Some shade will be needed in very hot places.

Asparagus: Propagation is usually by seed, though crowns may be obtained for transplanting. As it does not begin to crop until the third year, catch sowings of lettuce or radish or some other salad vegetable should be sown between the rows until space is needed by the asparagus.

The plant is a gross feeder and requires plenty of nutrient solution. The growing medium should be relatively loose in texture and well aerated.

Aubergine (Egg-plant): Garden eggs thrive best in a warm, dry place. Plenty of nutrient solutions will be needed, and a moist aggregate should always be maintained. To secure the most abundant fruits, the points of growth may be reduced in number by pinching out. When fruit is forming, to prevent the growth of leaves at the expense of the eggs, substitute plain water for nutrient solution every few days, returning to a standard fertilizing procedure when the fruits are well developed.

Beans: Many kinds of beans thrive in hydroponics, including broad, french, scarlet runner, lima, asparagus, velvet, soya, and other types. The cultivation of beans is not difficult. Generally speaking, these crops demand good irrigation but should never be allowed to become waterlogged. Aeration of the roots is essential. Different species need different degrees of warmth.

Beetroot: Sowing can take place in trays of sand, the young seedlings being transplanted into the primary containers when two to three inches high. As growth proceeds, a little extra aggregate should be placed around the roots. Beetroots are well adapted to hydroponics, and their light requirements are moderate.

Brassicas: These include broccoli and cauliflowers, cabbages, Brussel sprouts, pe-tsai or Chinese cabbage, red cabbage, kale, kohl-rabi, and turnips. All belong to the brassica group of vegetables. They are sensitive to lack of air in the growing medium but need frequent applications of nutrient solution. The drainage of the hydroponic unit must be kept in good working order when brassicas are grown. Cabbages, in particular, are gross feeders. Cauliflowers do best when they are unchecked and can make heads quickly.

Carrots: Carrots need quite a lot of attention, but once the seedlings have taken hold, good growth will result. The shorter varieties may penetrate better into the aggregate. Too coarse or heavy, a growing medium is not desirable for this crop.

Celery: For ease of sowing, the seeds may be mixed with a filler of sand. The plants need a fresh and partially shaded position and will require blanching before harvest.* This can be done by putting plastic collars on the stems. Proper aeration is essential for celery production in hydroponics.

Chicory: Seed can be sown in small boxes filled with sand or vermiculite, later on transplanting into the permanent growing container as convenient. For forcing chicory, the pots or troughs may be covered with inverted flower-pots or trays, but a better method is to lift the crowns and trim off any side-roots. The crowns should then be stacked in small boxes filled with aggregate, up to a depth of an inch or two. The growing medium must be kept just moist with a solution, and the whole unit put in complete darkness at a temperature of between 55 ° and 60 ° F. When the shoots attain a height of about seven inches, they may be cut off and used. It is essential to ensure that the blanching is thorough.

Chilies (Sweet and Hot Peppers): These need plenty of nutrient solution, but do not like a humid atmosphere. Light shading can be given in hot areas. Care must also be taken to provide protection against strong winds. In cold regions, cultivation should be inside the home or in greenhouses. Self-blanching varieties of celery are now available, but like the new tomato hybrid strains, the flavor may not always compare with the older, conventional types.

Cucumbers: Seeds may be grown in sandboxes or trays, and the young seedlings transplanted later into the main troughs or pots, making sure that there is no check to growth. Some support must be provided for the vines to climb up. The best day temperatures for cucumbers are generally 75 ° to 85 ° F, and they will stand over 100° F. Direct seeding is also quite practicable. For the proper development of female flowers, a program of terminal bud pinching is needed, unless the new varieties are grown. Cucumbers have very high water and solution requirements, needing frequent irrigation, relatively high atmospheric humidity, shelter from wind, shading during bright periods, and good general care.

Endive: This salad vegetable is natural to grow in soilless home gardens, being very similar to lettuce. For blanching, draw the leaves together and tie them up, so as to exclude light from the heart of the plant, or else cover the endives with inverted flower-pots, adjusted so that the air goes under the bottom.

Leeds: These are gross feeders. They will need ample amounts of nutrient solution. The stems can be blanched by putting on plastic collars. To prevent leeks from running to seed in hot conditions, cut the taproot with a sharp knife.

Lettuces: Lettuces should be grown quickly without any checks being permitted during development. Seed may be sown in sandboxes, and the young seedlings later transplanted to the primary containers. They can also be sown directly. A filler of fine sand can be added to the small seeds to ensure better distribution.

In hot places, lettuce plants need shading, and the troughs should be kept reasonably cool. It is also essential to see that no waterlogging occurs or else the lower leaves will rot, and the root crowns may be damaged. This salad is quite simple to raise in household hydroponic units.

Melons: The two main types are cantaloupes or musk melons and watermelons. The former like hot, dry conditions, with ample irrigation and arc susceptible to high air humidity. It is an excellent plan to place a small collar of damp-excluding material around the base of the plant's stem to keep it dry during growth. Watermelons also prefer dry and warm spots but are hardier than cantaloupes. Both types may need hand pollinating.

Onions: Onions need reasonably dry conditions. Dampness does not favor proper development. Spring onions may be treated like other salads, but for mature onions, it is best to stop any irrigation once they have attained full size and allow the tops to dry out. The bulbs can then be lifted for storage.

Peas: Both dwarf and tall types are available. Peas need good aeration, freedom from very humid conditions, and proper supports. In dry places, ample amounts of nutrient solution will be necessary. Seed may be sown directly in the containers.

Potatoes: It is essential to ensure that the tubers are provided with an adequate covering of aggregate, otherwise overheating in warm times and greening may occur. The plants can withstand hot conditions provided the root zone is kept fresh.

Proper aeration and drainage are vital, but potatoes also need ample supplies of nutrient solution.

To sow the seed tubers, simply make holes in the growing medium about three inches deep and drop in the potatoes, raking back the aggregate over them. In household hydroponics, the main interest in cultivating this crop will be to obtain better-tasting new potatoes.

Radishes: Radishes cannot be transplanted, but should be sown directly into the main troughs or pots, often as catch or intercrops. They are very quick growing. Proper shading is necessary for hot areas, or the plants may bolt.

Spinach: This crop needs ample applications of nutrient solution and a fresh, shady site. There are several kinds of spinach, but all usually are fairly quick maturing. Seed may be sown in trays of sand and subsequently transplanted to the primary hydroponic containers.

Sweet Potato: This crop needs a warm place with reasonably dry air conditions and shading from the hot sun. It is a moderate feeder.

Tomatoes: Tomatoes should be protected from strong wind and excessive humidity. The plant likes warm conditions, some shading, and proper aeration. Seed is best germinated in boxes or trays of sand and vermiculite, or other finer aggregates, and the young seedlings later moved to their permanent growing positions. The vines will require supports. Tomatoes become stunted if kept short of nutrient solution, so checks to growth should not be permitted. After the fourth truss, the main stem can be cut off, and a side shoot was taken on. This operation will restore pristine vigor to the plant. Suckers must be removed as they appear. The flowers of tomato plants can be syringed gently from time to time to encourage pollination, with better fruit formation. Tomatoes are probably the most widely grown vegetable crop amongst amateur hydroponicists.

Watercress: For the most effective growth of this crop, the solution should be allowed to flow continuously from an elevated tank or bucket into the aggregate.

As it passes out of the drainage holes, which are kept open all the time, it is collected in a sump or vessel and returned, at frequent intervals, to the overhead reservoir for recycling. Although the seed is sometimes used, cuttings from selected plants give the best results. The temperature for watercress should be maintained at from 60 ° to 65 ° F. The rate of flow of the liquid should not be too quick; a speed of about one mile an hour is quite adequate.

Yams: Deeper beds will be necessary for this crop. Shading, too, is essential. Care must be taken not to over-irrigate, or the tubers will rot. The yam seed should be planted about ten inches below the surface of the growing medium and supports provided for the vines to climb on.

FRUITS

Strawberries are a popular hydroponic garden fruit crop. Runners are most suitable for starting in the troughs or beds in the soilless unit. They may either be forced for fruiting during the first season or left in for three years; the best berries were then obtained in the second and third years. Much depends on the variety or type selected for growing. Care must be taken not to allow the plants to become waterlogged, nor should the leaves get saturated with a nutrient solution or rot may set in. Several other kinds of soft fruits can be grown in hydroponics. Very successful results have been secured with papayas or pawpaws, and pineapples which do well in greenhouses or conservatories. Grapes also thrive, given large deep pots. Vegetable crops mature earlier in soilless gardens than they do in soil. A sowing and planting program for the household should take account of this fact so as to ensure that a continuous supply of fresh produce is available at all seasons. In Florida, growers obtain hydroponic tomatoes seventy days after planting seedlings.

Californian experience shows that tomato fruits will be ready within sixty days or two months. Lettuces raised in hydroponics in England have produced good hearts a week or ten days earlier than similar sowings made in the soil. French beans and radishes produced in United States Army soilless gardens took only thirty-five days to mature.

High yields are characteristic of hydroponics. With due care and proper growing conditions, your tomato plants can yield up to thirty pounds of fruit each. In commercial units, it is common to get harvests of as much as two or three hundred tons per acre. Similar results have been secured with other crops. Seeds vary greatly, according to variety and quality, in the time they take to germinate. Never buy doubtful seeds from unreliable firms; it is a waste of money.

Always get fresh ones of guaranteed origin. Here are some average germination times for seeds in hydroponic home gardens:

Different varieties may be quicker or slower to germinate, while temperature plays an essential part in the length of time taken. The advice on seed packets should always be read carefully.

HERBS

These can be grown without difficulty in the kitchen, either on window ledges or by fixing up simple plastic tents just outside. If the tent is made from polythene sheets and designed in a lean-to style, it will be possible to allow hot air from the kitchen to flow through an open window or ventilator into the growing compartment, keeping it nice and warm in the wintertime. Suitable culinary species that do well in hydroponics include borage, chervil, chives, fennel, garlic, horseradish, marjoram, mint, parsley, rosemary, rue, sage, savory, tarragon, and thyme. In growing hydroponic vegetables, salads, fruits, and other edible plants in the home, the aim should be to secure as large a yield as quickly as possible from the smallest given area. With flowers, the object is frequently to obtain *massed bank* effects.

For foodstuffs, we usually like to have constant supplies, and one way of achieving this is to sow a succession of crops, having another ready to use as soon as a previous and earlier lot is exhausted. Careful planning and a little forethought are all that are needed to get such results. As mentioned before, the spacing between plants in hydroponics may be reduced, by as much as fifty percent, as long as light availability is assured. This also assists in the production of higher yields from limited space.

OTHER USES OF HYDROPONICS

Apart from the culture of flowers and vegetables, as well as fruits and herbs, for the home and household garden, there are a number of other ways in which hydroponics may come in handy for the family. In this chapter, several such uses are described, and instructions are given on their application.

GERMINATION NETS

Germination nets are quite useful for starting difficult seeds, which, as they grow, may be moved into permanent containers. The method of construction consists of taking a piece of ordinary cotton mosquito netting and dipping it into melted paraffin wax.

Candles can be used for the purpose. While still as hot as possible, the impregnated netting is tightly stretched over the top of an enamel-ware pan of convenient size and bound firmly beneath the marginal rim of the basin with a strong cord or string.

The nutrient solution is then poured into the pan until the surface of the liquid comes into contact with the bottom of the net. The seeds, which have been previously soaked between pieces of damp blotting paper to make them swell, are now sprinkled on the treated net, where they soon develop, being in constant contact with the solution, yet freely exposed to the air. After they have attained a few inches in height, they may be transferred to the pots or containers in the hydroponic garden.

INCREASING YOUR PLANTS

Although not everyone wants to propagate his or her own stock, other than seeds sown directly and then thinned out or transplanted, for those interested, here is a shortlist of the possibilities of increasing your soilless garden plants by simple means.

CUTTINGS

These are easy to root in aggregate. Always ensure that stem cutting is taken with a joint or node at the bottom. Remove the lower leaves and insert the stem half-way up into the growing medium. Stem, root, and leaf cuttings, as well as offsets, are frequently used, too, for specific plants. To assist root formation, rooting hormones may be bought from garden stores and prominent chemists.

To layer runners, simply peg them down at a convenient point half-an-inch into the aggregate.

Roots will form in due course. Bulbs, ferns, and some other types can be divided up to make more plants.

Bulbs form tiny bulblets naturally, while root clumps may be separated with care. Afterward, transplant the new specimens to their growing sites or containers.

AIR LAYERING

This method is utilized for large ornamental plants or those difficult to root. Make a cut in the main stem, about two-thirds through it, and keep this open with a piece of a matchstick. Wrap some moist sphagnum moss (available from garden stores or perhaps florists) around the open cut, enclose the whole in a piece of polythene sheeting and tie the ends to form a moisture-proof pack.

When roots start to develop inside, cut off below the pack, remove this, and plant the top stem in another pot. New shoots will soon arise from the old stump.

Many home gardeners using hydroponics, prefer to sow their seeds separately in small boxes filled with sand or sand and aggregate mixture, later transplanting them into the permanent containers, when they are about two inches high. In soilless culture, removal is simple, and there should be no check-in growth.

Cacti and succulents will do well in hydroponics, but always keep the aggregate slightly drier than for other plants, and never over-irrigate with a nutrient solution or you will get rotting of the stems and roots. Bulbs and corms, too, present no trouble; even a heap of moist aggregate on a saucer, regularly given solution, is adequate for them. Again, do not wet excessively and see that they get enough air. Fruit pips, planted like seeds, will often make pretty and attractive small household trees. Oranges, lemons, grapefruits, citrons, dates, apricots, avocado stones, and peaches are amongst the easiest to germinate.

Remember also kitchen herbs, useful for culinary purposes, and for scent and small flowers or blossoms.

Then there are root vegetable tops for greenery and similar miscellaneous attractions. Rampion and dandelions, as well as corn salad, are quite useful at times and are well suited to soilless household gardening.

A WORD OF WARNING

If you are given or buy, flowering plants or vegetable seedlings from nurseries, garden stores, or shops, rooted in soil or composts, never plant these directly in your hydroponic unit. To do so will cause a serious risk of disease and upset the scientific balance of the soilless garden. Also, very often, such specimens have been forced under unnatural conditions and will soon die when taken into the home. If, however, you must have bought-in or gift plants, first treat them in the following way before putting them in your hydroponic unit.

Remove the plant from its pot, if it is in one, by up-ending it, and place it with the soil still around its roots in a basin of clean water.

Let it soak for a while until the earth or compost is freed from the roots. Transfer it gently to another basin, leaving the wet soil behind and repeat the process until the roots are quite bare and clean. Be careful not to mistreat it or damage any roots.

Then rinse them gently under a slowly running tap to wash away any vestiges of soil or organic matter. When this has been done, dip the plant, roots and all, two or three times in a solution of water and formalin, available from general stores or chemists' shops, prepared by adding one fluid ounce of the disinfectant to every two and a half to three pints of water. (There are twenty fluid ounces in a pint of water.)

After such treatment has been completed, put the plant straight into its new hydroponic container, arranging the roots carefully, not in one bunch, but spread out and dribble the aggregate, previously moistened, with solution, around them.

Continue until the plant is standing firmly in place and see that the growing medium holds it up nicely. Smooth over the surface and water well with the nutrient liquid. Keep quite moist, but well aerated for a while when it should have accustomed itself to its new surroundings, then treat it in the standard way. But there can be no guarantee that such bought-in shop plants will do as well as your own grown seedlings because of the forcing that they have undergone before the sale. The same treatment can, of course, be applied to young seedlings or plants in boxes, germinated in soil or composts, and generally offered to the public in the spring time.

FLOATING RAFTS

If you have a pond or other stretch of water available, you can make hydroponic water gardens very cheaply. These are based on the floating gardens laid down by the Mogul emperors of India in Kashmir and the ancient chinampa of the Aztecs of Mexico. To make such a soilless garden, you need to build a small raft of light wood or similar material, pierced by a number of small holes.

Cover this platform with a six-inch layer of light aggregate, with boards around the sides to keep the growing medium from falling off. Fit wicks through the holes in the bottom, passing from the water below into the aggregate. Sow or plant the chosen flowers in the aggregate, apply solution regularly, and rely for additional moisture on water moving by capillary action along with the wicks. Periodic aeration is arranged by moving the raft or platform from time to time.

HANGING BASKETS

The hydroponic hanging basket consists of a trough or container which is slung from a system of overhead pulleys above a shallow basin full of nutrient solution. At regular intervals, the basket, which has a beautiful wire mesh base or else a number of inlet holes, to enable the nutrient solution to penetrate to the roots, is dipped into the liquid, thus irrigating and feeding the plants growing in the container. A light aggregate mixture should be put in the wire mesh trough to anchor the plants' roots in position.

ORCHIDS

To grow orchids, the best potting mixture is made up of equal parts of broken bricks, wood chippings, bark or shavings, and sphagnum moss. For terrestrial orchids, charcoal, small pieces of bone, and some coarse sand may be added. The plants are usually propagated by means of cuttings or divisions of the rootstock or pseudo-bulbs, although experts do raise them from seed. The germinating medium should be infected with the appropriate microscopic fungus mycelium, pure cultures of which are available. In the absence of the fungus, orchid seeds seldom germinate, or if they do, the seedlings will not thrive. Orchids should be given conditions that resemble those of the areas from which they originate, as far as temperature and humidity are concerned. This is usually done by growing them in greenhouses or conservatories.

MUSHROOMS

Mushrooms are lower plants or fungi and differ from the green or higher plants in the way in which they feed. They cannot, therefore, be grown in standard hydroponic growing media or by using reasonable nutrient solutions.

In small-scale laboratory work, mushrooms may be cultured in agar-agar jelly with added sugars, or in treated sawdust with some added oatmeal. This method would be too costly and troublesome for the housewife or amateur gardener. However, because mushrooms are a popular domestic crop, a simple method of soilless cultivation has been devised for home use.

The easiest way to grow mushrooms at home is to raise the crop in shallow wood troughs filled with a single growing medium.

This method is particularly suitable for householders who may wish to produce moderate quantities for family consumption.

The troughs can be kept without inconvenience inside rooms, in attics or cellars, or in garden sheds. By arranging the trays or troughs in tiers, or in series on shelves, it is possible to raise considerable quantities of mushrooms from comparatively small areas, while periodical sowings will ensure that a crop is available all the year-round.

Cheap growing troughs can be made from fruit boxes, of the kind used for packing peaches or grapes. The best dimensions are about three feet long by two feet wide by some six inches deep. Each box should be lined with polythene sheeting.

A gap of a foot should be allowed between troughs in tiers. Normally a trough with a growing space of six square feet will yield about seven pounds of mushrooms from every harvest period.

By nailing strips of wood to the ends of sets of boxes, it is quite simple to erect inexpensive tiers of mushroom troughs, and a number of such devices would form a convenient unit, depending on the space available and the production desired.

Once the troughs have been prepared, the next task is to make the growing medium. This is done as follows:

Take three hundred pounds of straw, place it on a large polythene sheet or on a concrete backyard, and wet it thoroughly with water. Leave this for two days. Then mix together one hundred pounds of sawdust, thirty pounds of bran, and enough water to moisten them.

Add nine pounds of ammonium sulfate, nine pounds of superphosphate, and four pounds of urea fertilizer. Make sure that all the nutrient salts are absorbed, and the whole lot is well mixed.

Now combine the straw with the sawdust, bran, and fertilizers and pile the entire mixture into a heap about four feet long by three feet wide by some three feet high.

Water well so that the material is quite wet, with some seepage from the bottom of the heap.

In two days, the temperature of the heap should rise to 140 ° F to 160 ° F.

On the sixth day, turn the material for the first time, by removing one foot of the compost from the sides and top of the heap, breaking up the remainder, and replacing so that the outer parts now form the inside, and the inner portions become the outside.

This procedure allows for gas exchange to occur. While turning, sprinkle ten pounds of calcium carbonate over the material. Apply more water, ensuring that no dry patches are left in a heap.

The temperature will rise again to about 150 ° F. On the tenth day turn for the second time, and then on the thirteenth day, give a further turn, sprinkling twelve-and-a-half pounds of gypsum powder (calcium sulfate) over the material. Reform the heap, as before, and make new turns on the sixteenth and the nineteenth days.

On the twentieth day, break down the heap, spreading it out on the polythene sheet or concrete and sprinkling half-a-pound of Lindane insecticide over the soilless compost to check any pests that may be present. The material should be dark brown in color, without any objectionable odor, and possessing ample moisture when pressed in the palm of the hand. This is the soilless growing medium, and it is now ready for use.

There will be enough to fill about a dozen troughs of the size already suggested for household use.

The soilless growing medium or substrate is placed in the troughs to a depth of some five inches and pressed down gently. If necessary, pieces of spawn each about the size of walnut are placed 2 Inches deep In the substrate and 10 Inches apart spray carefully with a little water to increase humidity. The temperature should rise to about 138° to 140 0 F. Maintain this heat for two days, then allow it to fall to 75 ° F by extra ventilation or inflow of fresh air.

Spawning may now be carried out. Pieces of spawn, each about the size of a walnut, should be inserted into holes two inches deep and ten inches apart in the growing medium. Grain spawn can be broadcast. The soilless compost around the pieces of spawn must be pressed firmly against them to give good contact, or the mycelium will not spread well.

Keep the temperature at from 70 0 to 75° F and certainly not below 65 ° F. Spray the troughs lightly with water daily. The mycelium will completely impregnate the growing medium or substrate in about two weeks, though occasionally it may take three. You will see the whitish thread-like growth permeating the soilless compost. As soon as this happens, the troughs should be covered with a layer of sand one-and-a-half inches deep. This job is called *casing*.

After casing, lower the temperature to 58 ° to 65 ° F, which is just right for mushroom production. Lightly spray the trays or boxes twice daily with water. The mushrooms should start appearing after fourteen days.

A good long thermometer is desirable for controlling mushroom growth. Keep a careful check on readings and use it regularly, inserting it right down into the growing medium. Mushrooms should be picked by twisting the roots gently, taking them out with the tops, disturbing the casing as little as possible. Any holes left should be filled immediately with fresh sand. Harvesting may last about two months.

When the bearing has finished, discard the old growing medium in the troughs or boxes and replace it with a new lot. Careful watering helps to secure good yields. During cropping periods, it is reasonable to spray the troughs lightly with water to prevent the casing from becoming hard and dry. Light and frequent applications are best, rather than heavy ones, at longer intervals. Proper ventilation is essential to replace the fresh air used by the plants. Mushrooms exhale large amounts of carbon dioxide.

Try to ventilate rooms and sheds for at least a quarter of an hour daily, but exclude any draughts. To keep troughs warm, it is practicable to use an electric heating cable or hot water pipes.

In mushroom growing with this soilless method, cleanliness is essential. Remove all stumps and pieces of stems left after picking. This prevents the growth of molds on troughs. Weekly dustings

The insecticide is helpful in keeping away pests like woodlice.

Mushrooms have high food value, and the flavor of homegrown soilless garden ones is excellent, far better than stale shop-bought lots.

BIG SCALE LAYOUTS

Once you have started a hydroponic home garden and become accustomed to the ease and simplicity of soilless plant growing using the simple method described in this book, you may well feel, if you have space, that you would like to extend your unit. In the case of indoor pots and troughs, this may be just a matter of multiplying the number of containers.

However, where circumstances permit, other possibilities exist. For example, one can frequently layout a small roof garden or use balconies and backyards or the edges of paths for exciting and profitable displays of hydroponics.

Even in winter, it is quite simple to use polythene sheeting to erect cheap covers, plant houses, cloches, and similar shelters for tender plants outside the house.

Glass-enclosed verandas protecting front doors or french windows also provide scope for tiers of attractively arranged containers. Of course, at cold times, it is necessary to fix up some form of heating, either by connection to the house installation or from electric heaters and paraffin stoves. Finally, for those with a greenhouse or conservatory already available, conversion to hydroponics has numerous advantages. In hot regions, the merits of garden plant houses that afford shade and cooling sites are well known.

LARGER CONTAINERS

Usually, it is not recommended in bigger layouts that troughs should exceed three to four feet in width, otherwise attending to plants becomes difficult, but they can be of any convenient length, either straight or curved, according to individual taste.

Usually, a depth of about six inches for most plants should be adhered to, except where particular species are to be cultured. Some root vegetables, ornamental shrubs, and fruit trees like papayas or bananas will need deeper containers.

With more significant troughs or beds, the drainage holes are generally placed at the sides, with narrow guttering to carry away seepage. For ease of solution application and watering, a system of pipes or a prefabricated irrigation unit, there are several on the market, advertisements for which appear in the gardening and farming journals, or just a hose can be employed.

Additional refinements include automatic nutrient applicators, with time clocks, heating cables, and other apparatus. But here you are getting close to semi-commercial work, which is outside the scope of this present book.

LAYOUTS

Convenient and well-designed layouts for hydroponic gardens of bigger size will help householders and amateurs to work more efficiently and economically. Save yourself as much time and labor as possible. There is no hard manual work in soilless culture, but you can still look for improvements in the arrangements for maintenance.

Proper planning in advance brings better returns. If you are interested in expanding your hydroponic unit into a sizeable installation for producing ample amounts of green food and flowers, perhaps surplus to your family needs, which you might sell part of, then organize it into the following sections.

- the propagation section

- the production section

- the service section

The propagation section is where you raise small plants for eventual transfer to the production department.

The latter is concerned with the growing and output of the crops themselves. The service section includes the labor and facilities that keep the project moving, that is your own sound right arm, plus possibly the help of your family, and the knowledge supplied by your learning.

Also, the tools that you use. As a general rule, the total area of the propagation section need not exceed three to four percent of that of the production section.

The propagation department should constitute a self-contained outfit situated near the main troughs or containers in a convenient spot. If possible, propagation boxes or trays should be raised above ground level.

This saves much time because it avoids any need for kneeling or stooping during the routine operations of sowing and caring for seedlings and other tasks. In the production section, any number of containers may go to make up a unit or installation. In large-scale work, the United States Army used as its standard growing unit a set of ten troughs, each three hundred feet long by three feet wide.

A ten-bed unit occupies approximately half-an-acre of ground, with paths and ancillary services and facilities. In greenhouses, troughs may be erected on existing staging or benches. Often five to ten beds are operated as one unit, with walks provided between the rows of troughs or containers.

For example, The Miami hydroponic gardens, in Florida, are based upon an original design of one hundred troughs, each a hundred feet long by three feet wide, with the nutrient solution flowing down a particular flume to irrigate the growing medium in these containers.

When hydroponic garden units are started in hot tropical localities, it is best to site the troughs or containers on a north-south axis.

This facilitates the effective shading of plants with overhead mats or canopies from the scorching rays of the noonday sun. In temperate colder places, the best use should be made of available light.

If strong winds are prevalent, the troughs or containers should be laid out, if possible, parallel to the direction of the wind. However, this need be no obstacle to hydroponic gardening in Colorado growers have to contend with winds of one hundred miles an hour at certain times of the year, but suffer no severe effects. The economy of space should always be aimed for. The walks between individual troughs need not be more than three feet in width, just enough to admit barrows and other appliances for harvesting, if the size of the unit justifies their use.

Whatever the shape or style of layout for larger installations is chosen, it should be tailored to fit the requirements of the householder or amateur gardener.

Use your ingenuity to make your own designs and improvise as much as you can. To save expense, employ whatever materials and equipment you may have on hand. Never waste money on buying costly devices and unnecessary apparatus. Any person can, with the exercise of a little imagination, design for him- or herself very pleasing and attractive layouts for soilless gardening at home.

In bigger hydroponic units, it is just as well to have a shed for storing your fertilizers and other equipment. There you can keep your hosepipes, watering cans, spare containers, seeds, weighing scales, tools, and any other items you use, all tidy and in good order, ready to find just as you require them.

Dry storage arrangements will be needed for fertilizers, insecticides, fungicides, and so on. Also, a table for the scales or balance and for writing your notes and records.

TESTING

As you become more advanced in working larger hydroponic units, you may like to do a little testing. Strictly speaking, these procedures are really designed for commercial operators engaged in large-scale production.

However, you can quite easily check on the quality of your nutrient solution as it is absorbed by the hydroponic plants. Hydroponics experts regard the balance of the liquid plant food or its pH, which is the alkalinity or acidity of the solution, as a primary factor in ensuring that flowers and vegetables grow well.

To test the pH, you will require what is called an indicator. This can be bought or ordered from a good chemist. Ask for a Universal indicator.

Having obtained a small bottle of indicator, you should proceed as follows.

About a quarter of an hour after irrigating or applying the nutrient solution to the growing medium in a hydroponic container, make a hole some four or five inches deep in the aggregate, near a plant.

You should find some moisture at the bottom of this hole. Using a medicine dropper, suck up a little of the liquid and put it on a clean white saucer. Before doing this, take care to wash the dropper in clean distilled water thoroughly and also put on clean well-washed rubber gloves.

Now with another dropper, also washed in distilled water, take a tiny portion of the Universal indicator from its bottle and squeeze it into the solution you have taken from the growing medium. Observe the colour change. If the result is a red or orange liquid in the saucer, the solution is too acid. A green or greenish-blue color means that it is too alkaline. The right color is yellow to yellow-green. That will give you a pH reaction of about neutral or in technical terms 6.0 to 6.5. Actually, anything from about 5.0 to 7.0 is tolerable, but the most satisfactory reaction is in the region of 6.0.

Further information on the testing procedure and how to control the pH reaction will be found in more advanced books on hydroponics. If you are using a proprietary formula or nutrient mixture, you should get in touch with its manufacturer to ask for comments and advice.

TIPS AND TRICKS

This chapter contains a variety of information that may be useful to beginners in soilless gardening, ranging from advice on different aspects of the system, weights and measures, a list of books for further reading, and miscellaneous notes for housewives and amateurs.

HYDROPONICS AS A NATURAL SYSTEM OF PLANT GROWING

From time to time, you may hear criticisms that soilless cultivation is an unnatural way of nourishing and growing plants. Now, it is entirely right that artificial fertilizers are often injurious in soil gardening and farming because their use alone and unrestrictedly can destroy the land and eventually result in severe erosion. Therefore, in soil cultivation, mineral salts should always be applied in conjunction with organic manures. But in soilless gardening, there is no soil to damage, and therefore this criticism is in no way valid.

The growing media in which hydroponic plants thrive and develop is inert, durable, and not subject to erosion or destruction. As we know, plants derive the more significant part of their food from the air, but they cannot exist without supplies of mineral elements and water.

When growing in the earth, crops absorb these from the soil through their roots. Some of the mineral salts originate from the weathering of rock fragments, but nitrogen, one of the most important, is produced mainly from dead vegetation and animal remains.

Of course, no organic manures can be directly absorbed by the roots of higher plants, but must first be broken down by bacterial action to ammonium compounds and finally to nitrates. Only then do they become of any value to the feeding of the plants. Thus when farmyard manure or compost are applied to the soil, they have first to be converted in it to an 'artificial' fertilizer before they can become available for crop nourishment.

Green plants can only take up inorganic salts in solution with water; they cannot 'cat' any organic matter or humus. In hydroponics, we simply give the mineral salts direct, no process of change has to occur, and consequently, the needs of the plants are met immediately.

There is nothing unusual about this because chemical fertilizers arc basically the products of normal and natural substances.

Hydroponics is in no way unnatural, it merely cuts out an unnecessary part of the life cycle, avoiding the lengthy process needed to change manure or organic material into an inorganic salt. Growth is also speeded up as each plant is assured of maximum sustenance at all times. The soilless method is simply quicker and better, under the right circumstances than soil cultivation and a lot cleaner and more natural too.

FOOD VALUES

The nutritional value of hydroponic vegetables and fruits is satisfactory. Analyses have shown that the mineral and vitamin contents are quite reasonable.

In fact, it is possible to grow tomatoes containing extra calcium for feeding to babies and more iron for invalids. The same applies to other produce. Wheat raised in soilless gardens has been found to be better for bread making. The flavor of hydroponic greenstuff and fruits is excellent, and the author has never found any person who was not impressed by the taste and quality of the produce from well-run soilless units.

NEMATODES

Perhaps you have seen these small pests, also called eelworms, in garden soil or composts. Occasionally, they have been known to occur in hydroponic growing media due to outside infection very often caused by planting seedlings with old soil still around their roots, bought from shops or nurseries, in soilless garden containers.

Nematodes can damage plants. These pests belong to many different strains and species. Probably the worst is the root-knot eelworms, which make galls on the plants' roots and can cause the death of crops.

Eelworm attacks soon result in wilting and stunting of plants. It is quite easy to see the tiny white or brownish spherical bodies of the nematodes in the aggregate and on the roots. One of the best ways of guarding against infection, if you are working with larger boxes or troughs, is to keep a few African marigolds growing at intervals in the hydroponic containers.

The roots of these plants release a secretion which repels eelworms so that they will not remain in the soilless beds. The leaves also give off a powerful odor. However, the effects of the marigolds will not harm your ordinary flowers and vegetables. This procedure is based on the principle of biological control, a system now finding many supporters in horticulture. It is cheaper, and frequently better than using massive doses of pesticides, though you can, of course, obtain products to clear severe infestations of nematodes.

SPROUTED GREEN FOOD

Make one of these devices and scatter the bean or pea seeds over the netting. Cover with a piece of damp blotting paper or thin cloth for a few days until the seeds have begun to germinate. Then remove the covering and keep the unit in a warm place, but not in too direct light. After a week or two, the tiny seedlings will develop into delicious sprouts with ample, tasty, green tops. You can pick them at whatever stage you like, just when the sprouts are the height that suits your culinary requirements. Such green sprouts go very well in salads, particularly at times when other produce is short.

SOILLESS COMPOST

These days there is much talk about soilless composts. Just what are these? Compost is a fertilizing mixture, the word being derived from the Latin term *componere*, to put together.

To make compost for regular garden use, we usually mix up some soil, organic matter, manures, and a *starter* to activate the materials. In due course, these will heat up and mature into compost.

Now, a soilless compost, of course, contains no soil, being usually made from mixtures of sand and peat moss, with some added fertilizers.

It does not need any heating process. This material can be used to grow plants in for limited periods, but as soon as the nutrients in it are exhausted, it has no further feeding value, and the plants growing in it will gradually weaken and die. Soilless composts may be considered a part of gardening without soil, but they are not pure hydroponics.

For one thing, they provide no permanent source of nourishment for plants. Then, they are frequently ill-balanced, and gardeners cannot control the feeding of plants growing in containers filled with them. Such composts bear no comparison with the advantages of inert aggregates for maximum production.

All the same, you may find a use for soilless composts from time to time for raising seedlings and occasionally for house plants or ring culture. Never add any manures or other dirty organic materials to your soilless composts.

Use only sand and peat, with some added nutrients, and remember their life is strictly limited to only a few weeks or months. Then you must throw your soilless compost away and replace it with a freshly made up one. Here is a good formula for making soilless compost

- Fine sand

- Peatmoss (preferably) sterilized with boiling water and clean)

- Hoof and horn meal

- Superphosphate

- Ground chalk

- Ground magnesium limestone

- Potassium sulfate

- Potassium nitrate

You can use empty one-gallon petrol or oil can, with the top cut off, to measure out the sand and peat moss. Mix them well together. Then weigh out the fertilizers separately on a kitchen scale or balance and mix them all together. Sprinkle the mixture over the sand and peat moss that you have already mixed up. A further thorough mixing must then be done to produce compost with all the ingredients evenly distributed throughout it.

After placing the soilless compost in the pots or boxes, water carefully, so that it is just damp. Over-watering should be avoided, or the plants may suffer from mildew and lack of aeration.

Note. The fertilizers mentioned above may be obtained from agricultural and horticultural merchants, at some big chemists, in shops of the general store type, garden centers, and nurseries in many cases.

OTHER HYDROPONICS METHODS

In this chapter, brief descriptions are given of various other methods of hydroponics. The idea is to explain concisely how these functions, so that the beginner, after having become familiar with soilless gardening by using the simple method described in the central parts of this book, may later go in for more elaborate techniques if he or she so desires. Not all of us want to indulge in what can be slightly ambitious projects, but there is no doubt that once you have been bitten by the hydroponic 'bug,' you may well feel impelled to exploit the system to the full. There is something fascinating and very satisfying in utilizing these resources of science and technology; today is known as modern biology to produce better plants under strictly controlled conditions.

Even on a small scale in the home, backyard, or garden, you can employ quite elaborate methods of soilless cultivation to raise crops of outstanding quality by means of different apparatus and contrasting sets of equipment.

All methods of hydroponics conform to the same basic system. Soilless cultivation is a branch of horticulture, which draws upon agricultural chemistry, engineering, plant physiology, and related sciences to achieve optimum results. For the sake of convenience, the different hydroponic methods or techniques may be classified under four main headings: water or solution culture, sand culture, aggregate culture, and miscellaneous practices.

Each method contains within its limits a number of sub-divisions, generally intended to suit specific areas or purposes. The choice of a particular method depends typically on factors such as climate, place, availability of requisites, and costs. There is no 'best' method of hydroponics as such; what matters is whether the technique selected meets the requirements of those using it and the environmental circumstances.

The simple method that has been recommended for beginners in the preceding chapters of this book falls under the heading a mixed aggregate culture, combining the advantages, as well, of sand and solution culture. It is well suited for use in the home and can be operated by amateurs and housewives without difficulty.

Moreover, no fancy equipment is required, while results can be just as good as those secured in far more sophisticated methods. So don't rush to set up advanced and possibly costly installations, until you are sure you really want them or feel quite confident of handling them to full advantage. Let us now discuss the different methods in turn.

WATER OR SOLUTION CULTURE

Plants of all kinds have been grown successfully in water or solution culture, and the method has been employed for many years in laboratory tests. On a larger scale, it has given good results in commercial work.

This is quite easy to set up for home use, especially for the growth of ornamentals in individual containers. Any kind of glass or porcelain vessel may be employed, but the best receptacles are probably ordinary wide-mouthed fruit jars of one or two pints capacity. Fiat corks, with a hole in them, or wads of non-absorbent cotton, should be fixed firmly in the necks to hold the plants in position. The roots are submerged in the nutrient solution, which fills the jar to within about two inches of its top, leaving a small space for air. Clear glass containers should be covered with dark paper to exclude light from the roots. To provide aeration, it is necessary to remove the plants from the jars every few days and then shake the solution vigorously or else to blow air into it with a bicycle pump or by mouth through a tube.

CONTINUOUS FLOW

With this method, the solution is placed in a vessel attached to a small reservoir by a siphon tube. An outlet pipe, operating as another siphon, extends from a point near the bottom of the jar to a second overflow vessel, so creating a continuous flow of solution through the whole apparatus. By raising or lowering the end of the outlet, it is possible to vary the level of the nutrient solution around the plant's roots. At intervals, the air is blown into the solution.

Virtually the same apparatus is used for the drip technique as for the continuous-flow arrangement. However, to avoid the need for blowing or pumping air into the solution a gap of about four inches is left between the end of the siphon tube which takes the nutrient solution from the reservoir to the container, and the funnels that receive it at the top of the container or jar. By careful adjustment of the siphon, the solution can be made to drip out. Each drop collects some air as it crosses the space between the end of the tube and the funnel. This bubble of air is carried down into the region of the plant's roots, which grow in the solution in the vessel.

SWISS VESSEL

Here we have an egg-shaped vessel. In its upper quarter, there is a small detachable tray filled with stone chips resting on a wire grid. The root crown of the plant is supported on this grid, while the stem protrudes upwards into the air. Removable lids with a hole in the cover the tops of such vessels. The plant's roots descend through the mesh of the grid into the solution, which is poured into the pot or vessel. Several sizes are available or can be made up entirely.

WICKS

Cylindrical lamp wicks, as well as those made from glass wool, arc used in hydroponics to irrigate the roots of plants growing on trays set above containers filled with solution. In this case, the roots do not usually descend down through the mesh into the liquid to obtain food, but instead, the nutrient solution passes up the wick to the roots, and so keeps them moist and nourished.

GERICKE

This has developed from the original technique utilized by Dr. Gericke in California. It has given excellent results under different conditions and is sometimes called tank culture. However, it may be a little challenging to handle for beginners living in damp localities. Nevertheless, the Gericke style of hydroponic units is being employed in Poland today for commercial production. Waterproof basins, troughs, or tanks are needed to hold the nutrient solution. A wire grid fits over the top of each container and serves as a support for the plants, whose roots descend through the mesh into the liquid below.

A covering of some material such as wood-wool, coarse peat moss, shavings, or rough-dry hay or any other form of litter, is placed on the grid.

This excludes light from the nutrient solution in the container and provides additional support and protection for the crops. Aeration of the root system is ensured by adjusting the level of the solution, so as to leave an air space between its surface and the base of the wire grid. In specific units, the air is blown through the liquid plant food from time to time, or extra ventilating holes are made in the sides of the troughs just above the solution surface and below the bottom of the wire grids. To operate a water or solution culture unit of the Gericke type, the following information will be helpful.

The tank or container may be of any convenient size as long as it is easy to reach all parts. Waterproof materials are necessary for construction. The wire mesh framework should be supplied with litter bed, peat moss or wood wool, wire frame, air space nutrient solution, brackets, or catches to keep it about two or three inches from the top of the tank. A low guard around it will serve to keep the litter in position.

Whatever material you use to form this litter bed should be spread carefully over the grid and be of even composition and quite level. Sow the seeds on the litter and sprinkle a little extra material over them to supply some covering.

Then moisten the litter with a nutrient solution. Meanwhile, you will have poured enough solution into the tank below to reach up just to the bottom of the wire grid. This will ascend into the litter bed by capillary attraction and keep it constantly damp, as well as providing the source of permanent nourishment for the plants.

If the material is allowed to dry out, the plants will, of course, die. Small seedlings can be set in the litter, if desired, with their roots positioned close to the top of the mesh. In time, these will grow downwards into the solution below. As soon as whatever plants you have selected are a few inches high, you will find that their roots are passing through the grid and feeding on the liquid in the tank.

When this happens, lower the level of the solution so that only about half of the length of the developing roots is immersed. The idea is to provide enough air for healthy growth. Continue to do this as the size of the roots increases. Remember that this aeration factor is very critical. Lack of air is indicated by a yellowing of the plants leaves. It is necessary to keep the tanks covered with the grid and litter bed, but from time to time, you can lift the tops off to inspect the roots and solution below.

Blowing air into the solution occasionally helps. If holes have been provided in the sides of the containers, it is quite simple to glance through them to check the level or insert the tube of a bicycle pump there.

Once the tank has been filled with nutrient solution, you will need to top it up periodically with a little plain water and every fortnight empty it and refill with another batch of the solution, freshly made up. Sometimes, the solution for water culture can be stronger than for mixed aggregate or sand methods. To prepare this, add up to fifty percent more fertilizer salts to given quantities of water. Very much depends on the growth and appearance of the plants, which you will have to judge in the light of your experience.

SAND CULTURE

There are several variations of this method. In its simplest form, it resembles very much the technique of mixed aggregate culture already recommended in this book for beginners. However, sand culture, which is using sand alone as a growing medium, suffers from the drawback that it may be liable to waterlog and retention of excess moisture in beds or containers. In practice, receptacles for sand culture need not be well waterproofed, and indeed beds or troughs can be laid down on any hard surface or sunk into the ground. This, of course, results in some loss of water and nutrients and higher ultimate costs.

Surface watering: Frequently termed the slop technique, here we have plants grown in improvised beds or receptacles containing sand only, the solution being applied by can, pipe, or irrigation system to the surface of the material. Free drainage is permitted. Surface watering is effortless to operate, but, as already stated, it is wasteful of water and nutrients. In addition, it can be very messy in the home.

New Jersey: Essentially, the New Jersey techniques involve the use of an elevated tank from which the nutrient solution flows by gravity into the troughs filled with sand. Alternatively, it may be pumped several times a day to give similar results. In household work, hydroponic window boxes with a sand growing medium can be fitted with rubber squeeze bulbs and check valves, so making periodic manual irrigation possible. Another way is to spray the nutrient solution over the surface of the containers or pots in the soilless garden. In the Middle West of the United States, large scale hydroponics operated by automatic solution flow into sand beds is known as the Withrow technique.

Automatic Dilution Surface Watering: A concentrated solution is diluted to the required strength and then applied to the sand-filled troughs or containers by means of a system of sprays. A low-pressure water-main is needed, in addition to an injection pump and reservoir. To secure improved drainage, a layer of small gravel can be placed at the bottom of the containers underneath the sand.

Wic: A double pot arrangement is used. The upper receptacle consists of an ordinary flower pot filled with coarse sand and fitted with a glass wool wick, which passes through the bottom hole into the lower container.

This second vessel is watertight and holds the nutrient solution. The wick, which is divided at the top and branches out to facilitate adequate distribution of the solution in the root zone, draws up the liquid plant food from the lower to the upper pot.

Drip: Using a feed line or thin tube, diluted nutrient solution contained in a tank elevated above the level of the plants' growing receptacles, is allowed to drip continuously on to the sand medium. The nutrient solution percolates through the sand, is collected in another tank, and pumped back to the original tank or reservoir at intervals. For drip techniques, a waterproof container is necessary.

Continous-flow: The hydroponic container should be placed on a stand, under which is kept a basin or bowl to receive any seepage or drr.inage. From a nearby tank, a siphon and feed pipe maintain a continuous flow of nutrient solution onto the surface of the sand in the trough or pot. This technique is quite handy for household use.

Modified Slop: The nutrient solution is pumped through devices that flood the surface of the sand beds or troughs at regular intervals. It then percolates down through the growing medium and eventually drains back into an underground tank for storage. The technique may not be very suitable for salad crops, such as lettuce, which do not tolerate large amounts of free liquid around their root crowns.

AGGREGATE CULTURE

Numerous different growing media can be used in aggregate culture. These include gravel, broken-up bricks, washed cinders, clinkers, stone chips, leca, pebbles, vermiculite, and other substances of inert nature. Mixtures of two or more are also practicable. Aggregate culture is much employed in commercial hydroponics in all parts of the world. All the same, despite its complications, the scientifically minded beginner can quite easily construct smaller soilless gardens, using a technique of semi-automatic aggregate culture in the home or amateur garden.

Sub-irrigation: Watertight troughs or containers are necessary, and these are filled with gravel or some other hard inert medium. The beds must be flooded periodically with nutrient solution and then allowed to drain. For irrigation, direct-feed and gravity-feed apparatus may be satisfactory. Sometimes, cascade troughs or tanks are built, so that the solution flows from one to the next before discharging.

Troughs or beds may be built of any suitable materials. A reservoir, a sump, and a pump will be required. The entire technique is automatic, and by virtue of the ebb and flow of the nutrient solution, the roots of the plants are well aerated. Sub-irrigation is a very economical way of feeding crops. There should be no loss of water and nutrients. In addition, the grower has complete control over the plants.

On the other hand, the apparatus can be expensive to install, and the units rather complicated to construct. Nevertheless, in large scale commercial soilless gardens, money is saved on labor and maintenance.

Flumel This device consists of a long, curved, artificial channel down which the nutrient solution is directed, so that each trough or bed may receive a correct proportion of the liquid as the flat sweeps by. Flumes are employed in commercial hydroponics, particularly in Florida.

Bucket and Gravity Feed: Usually a small trough is constructed about three feet above ground level, with a taller post secured to the rear for hooking a bucket on to, or if a larger unit is contemplated, then with an overhead rail to which a series of buckets may be attached.

A hose is joined at one end to the bottom of the bucket and at the other to the base of the side-wall of the trough containing the growing medium.

When the bucket has been filled with the nutrient solution and raised to the elevated position, the liquid plant food flows down into the trough or bed, irrigating the roots. By dropping the bucket to the ground level, drainage of the growing medium is accomplished by the outward and backward flow of the solution.

This method is quite convenient for householders but needs frequent attention.

Compressed Air Design: The nutrient solution is contained in large drums, placed at ground level. From there, it is driven into the troughs or beds by portable air compressors. Drainage is by normal gravity flow.

Wic Devices: These can be utilized in small aggregate cultures and are quite valuable for seedling propagation work. Basins containing the nutrient solution are located underneath the troughs filled with gravel or other growing media, which are kept moist by the capillary action of the wicks. These run from the basins up into the aggregate through holes in the bases of the troughs. They are fitted so as to draw a constant supply of nutrient solution from the basins to the growing media.

Bengal: This technique was developed initially at the Government of West Bengal's experimental station near Darjeeling in India, and has given excellent results subsequently in many other countries. Troughs or containers made from any suitable material are used. These are filled with a reasonably coarse aggregate, made up of a mixture of gravel or stone chips and sand. The nutrient formulae are applied, generally in the dry state, to the troughs and then immediately watered into solution with sprays or hosepipes. By sprinkling the fertilizer salts evenly between the rows of plants, proper distribution of the nutrients is assured.

Typical irrigation facilities are provided through systems of pipes in large units. The aggregate in the containers must be kept consistently moist.

In large installations, machinery is used for spreading the formula. The technique is simple to maintain, easy to operate, and cheap to install. It is also adaptable to widely contrasting conditions and can be very useful for amateur gardeners and householders as well as for commercial production.

Vermiculite Technique: When vermiculite aggregate is employed in hydroponics, care has to be taken that its water-holding properties do not give rise to an excess of moisture in the containers.

Vermiculite may retain supplies of available water or solution for hydroponic plant growth for up to ten times as long as some other aggregates do. It is very light and easy to transport. In general, there are advantages to mixing vermiculite with sand or finer media.

It is not always recommended for sub-irrigation work but does very well in slop or other cultures.

Aggregate Hydroponic Units for the Home: Here are some suggestions for operating a small aggregate hydroponic unit in the home or backyard.

A waterproof container must be used. Plastic, enamel, or other troughs are quite suitable, but if a metal receptacle is chosen, be sure to paint it first with good quality varnish or paint, but not tar.

You can aim to provide a supply of nutrient solution through a hole or aperture in the bottom of the trough or container for the plants in such a way as to ensure that all the aggregate inside is well and thoroughly moistened.

To achieve this, lay a small perforated pipe or rose in the bottom of the container, connecting this to the hole. A hose fitting should be fixed in the hole so that the pipe or rose can be attached to it on the inside.

At the same time, this will enable you to fix a further piece of piping to the outside of the container. This pipe, in turn, should be connected to a bucket or tank, also having an aperture at its base, with a similar hose fitting in it. To feed the plants, fill the vessel with a nutrient solution and site above the growth container.

The solution flows down into the aggregate by gravity, gradually permeating all the growing medium. To drain it off, change the position of the vessel or bucket to a lower situation so that the liquid plant food runs out of the trough or container back into the vessel.

Alternatively, you can have a fixed tank and feed by gravity, allowing the solution later to drain off into a sump, then pumping it back again into the elevated reservoir or storage tank. Feeding with a nutrient solution is best done early in the morning so that it can remain in the trough or container during the daytime.

In the evening, start the drainage process, permitting this to go on all night. The cycle can be repeated continuously, adding a fresh solution to the tank or bucket as required. The object is to keep the growing medium damp and fit for plant development. In preparing a container for aggregate culture, using this sub-irrigation technique, take care to ensure that you make the growing medium of small pebbles, washed cinders, gravel, or other clean inert material.

Do not use fine sand or ashes. The perforated pipe or rose at the bottom of the container should be laid flat on the base and covered with a sheet of glass, plastic, or other similar material, to prevent the aggregate being washed out through the perforations in the pipe.

It must, in fact, be screened from the material inside the container. The perforated pipe or rose can be made of copper, plastic, or rubber, with holes about one-eighth of an inch in diameter, pierced into it at short intervals, so that the nutrient solution can flow freely, but not too quickly, out of the pipe into the growth container. The container should be at least six inches deep, to allow ample root space. It is better to coil the pipe at the bottom of the container to get more even distribution of the nutrient solution.

Deeper containers or troughs may be needed for some plants. Standard nutrient solutions can be employed for plant feeding. In larger units, hydroponics may find it worthwhile to invest in a small electrical pump connected to a timer switch, which automatically supplies the container with the solution at regular intervals.

NOTES